Yarn Formation

Yarn Formation

Naresh Grover

RANDOM

RANDOM PUBLICATIONS

NEW DELHI (INDIA)

Yarn Formation

ISBN 978-93-5111-650-9

Published in 2015 in India by

RANDOM PUBLICATIONS

4376-A/4B, Gali Murari Lal, Ansari Road
New Delhi-110 002
Phone : +9111-43580356, 011-23289044, 011-43142548
e-mail: sales@randompublications.com,
info@randompublications.com, randomexports@gmail.com

Reprinted 2026

Type Setting by : Friends Media, Delhi-110089

Preface

The thickness of yarn is measured as Counts. Yarns are made in different counts like 2's, 4's, 10's, 16's, 20's, 24's, 25's, 30's, 34's, 36's, 38's, 40's, 60's, 80's, 100's, etc. We can consider like this. 0's counts are cotton fiber. 20's counts yarn is thicker than 24's yarn. Likewise 30's yarn is thicker than 34's. So when the yarn counts are increasing, the thickness becomes lesser. We can see, the higher the counts, the lesser the thickness. Yarn prices based on the thickness. Price of 20's yarn is lesser than 24's. Price of 30's yarn is lesser than 34's yarn. We must know, higher the counts, higher the prices. There are 2 qualities of yarn. Combed and Carded. The Fibre formation process includes change in shape, structure and properties of the thermoplastic polymer. The polymer pellets or granules are fed into an extruder where, through heating, their melting temperature is exceeded. The polymeric melt is then transported, under pressure, to the spinneret. Yarn formation methods were originally developed for spinning of natural fibres including cotton, linen, wool and silk. By and large the book is intended to meet the needs of students, scholars, industrialist and must for the shelf of every library."

I would like to thank my team for standing beside me throughout [illegible] career and writing this book. My special thanks go to "Random Pu[illegible] who have published the book.

– Nare[illegible]

...t my
...plications"

...sh Grover

Contents

1

Tools and Materials

KNITTING NEEDLE

A knitting needle or knitting pin is a tool in hand-knitting to produce knitted fabrics. They generally have a long shaft and taper at their end, but they are not nearly as sharp as sewing needles. Their purpose is two-fold. The long shaft *holds* the active (unsecured) stitches of the fabric, to prevent them from unravelling, whereas the tapered ends are used to *form new stitches*. Most commonly, a new stitch is formed by inserting the tapered end through an active stitch, catching a loop (also called a bight) of fresh yarn and drawing it through the stitch; this secures the initial stitch and forms a new active stitch in its place. In specialized forms of knitting the needle may be passed between active stitches being held on another needle, or indeed between/through inactive stitches that have been knit previously.

The size of a needle is described first by its diameter and secondly by its length. The size of the new stitch is determined in large part by the diameter of the knitting needle used to form it, because that affects the length of the yarn-loop drawn through the previous stitch. Thus, large stitches can be made with large needles, whereas fine knitting requires fine needles. In most cases, the knitting needles being used in hand-knitting are of the same diameter; however, in uneven knitting, needles of different sizes may be used. Larger stitches may also be made by wrapping the yarn more than once around the needles with every stitch. The length of a needle determines how many stitches it can hold at once; for example, very large projects such as a shawl with hundreds of stitches might require a longer needle than a small project such as a scarf or bootie. Various sizing systems for needles are in common use.

TYPES

Single-pointed Needles

The most widely recognized form of needle is the single-pointed needle. It is a slender, straight stick tapered to a point at one end, with a knob at the other end to prevent stitches from slipping off. Such needles are always used in pairs and are usually 10-16 inches (254-406 mm) long but, due to the compressibility of knitted fabrics, may be used to knit pieces significantly wider. The knitting of new stitches occurs only at the tapered ends. Fictional depictions of knitting in movies, television programmes, animation, and comic strips almost always show knitting done on straight needles. Both Wallace and Gromit and Monty Python, for example, show this type of knitting.

Double-pointed Needles

The oldest type of needle is the straight double-pointed needle. Double-pointed needles are tapered at both ends, which allows them to be knit from either end. They are typically used (and sold) in sets of four and five, and are commonly used for circular knitting. Since the invention of the circular needle, they have been most commonly used to knit smaller tube-shaped pieces such as sleeves, collars, and socks. Usually two needles are active while the others hold the remaining stitches. Double-pointed needles are somewhat shorter than single-pointed or circular needles, and are usually used in the 13-20 cm length range, although they are also made longer.

Double-pointed needles are depicted in a number of 14th-century oil paintings, typically called *Knitting Madonnas*, depicting Mary knitting with double-pointed needles (Rutt, 2003).

A cable needle is a special type of double-pointed needle that is typically very short and used to hold a very small number of stitches temporarily while the knitter is forming a cable pattern. They are often U-shaped, or have a U-shaped bend, to keep the held stitches from falling off while the primary needle is being used.

Circular Needles

The first US patent for a circular needle was issued in 1918, although in Europe they may have been used a little earlier. Circulars are composed of two pointed, straight tips connected by a flexible cable and may be used for both knitting flat or knitting in the round. The two tapered ends, typically 4-5 inches (10.5-13 cm) long, are rigid, allowing for easy knitting, and are connected by the flexible strand (usually made of nylon or coated wire). The tips may be permanently connected to the cable and made in overall lengths from 9 inches (23 cm) to 60 inches (150 cm) or composed of cables and interchangeable tips. This allows various lengths and diameters to be combined into many different sizes of needles, allowing for a great variety of needs to be met by a relatively few component pieces. The ability to work from either end of one needle is convenient in several types of knitting, such as slip-stitch versions of double knitting.

In using circulars to knit flat pieces of fabric the two ends are used just as two separate needles would be. The knitter holds one tip in each hand and knits straight across the width of the fabric, turns the work, and knits or purls back the other way. Using circular needles has some advantages, for example, the weight of the fabric is more evenly distributed, therefore less taxing, on the arms and wrists of the knitter and, the length of the cable may be longer than would be practical with rigid needles since the cable and fabric rest in the lap of the knitter rather than extending straight out past the arms.

The lack of a purl row in stockinette stitch, since in the round (commonly referred to as ITR) knitting is all done using the knit stitch, is often perceived to be one of the greatest benefits of ITR. Knitting ITR with circulars is done in a spiral, the same way as using double-pointed needles (usually called DPNs). Additionally, circulars eliminate the need to continually switch from one needle to the next, and there is no possibility of stitches falling off the back end of the needles, as may happen when using DPNs. Much larger tubes may be knit ITR, too, helping items to be completed more quickly.

Construction of garments such as sweaters may be greatly simplified when knitting ITR, since the finishing steps of sewing a back, two fronts, and two sleeves of a sweater together may be almost entirely eliminated in neck down ITR knitting.

Numerous techniques have been devised for the production of narrow tubular knitting on circular needles. One common method is to use two needles in place of the four or five double-pointed needles traditionally used, while a newer technique is to use one circular needle that is significantly longer than the circumference of the item being knitted. This technique is known as Magic Loop and has recently become a popular method of producing tubular knitting, as only one needle is required.

THE GUINNESS WORLD RECORD FOR KNITTING WITH THE LARGEST KNITTING NEEDLES

The current holder of this title is Julia Hopson of Penzance in Cornwall. Julia knitted a tension square of ten stitches and ten rows in stocking stitch using knitting needles that were 6.5cm in diameter and 3.5 metres long.

NEEDLE MATERIALS

In addition to common wood and metal needles, antique knitting needles were sometimes made from tortoiseshell, ivory and walrus tusks; these materials are now banned due to their impact on endangered species, and needles made from them are virtually impossible to find. Modern knitting needles are made of bamboo, aluminium, steel, wood, plastic, glass, casein and carbon fibers.

NEEDLE STORAGE

A tall, cylindrical container with padding on the bottom to keep the points sharp can store straight needles neatly. Fabric or plastic cases similar to cosmetic bags or a chef's knife bag allow straight needles to be stored together yet separated by size, then rolled to maximize space. Circular needles may be stored with the cables coiled in cases made specifically for this purpose or hung dangling from a hanger device with cables straight. If older circulars with the nylon or plastic cables are coiled for storage it may be necessary to soak them in hot water for a few minutes to get them to uncoil and relax for ease of use.

Most recently manufactured cables eliminate this problem and may be stored coiled without any difficulty. Care must be taken not to kink the metal cables of older circulars, as these kinks will not come out and may damage or snag yarn as it is knit.

NEEDLE GAUGE

A needle gauge makes it possible to determine the size of a knitting needle. Some may also be used to gauge the size of crochet hooks. Most needles come with the size written on the them, but with use and time, the label often wears off, and many needles (like double-pointed needles) tend not to be labelled.

Needle gauges can be made of any material, but are often made of metal and plastic. They tend to be about 3 by 5 inches. There are holes of various sizes through which the needles are passed to determine which hole they fit best, and often a ruler along the edge for determining the tension (also called gauge) of a sample.

KNITTING NEEDLE CAP

A knitting needle cap, also known as a point protector, is a cover placed on the tip of a knitting needle that is being used for a project that is resting. The cap prevents stitches from coming off the ends of needles. There are knitting needle caps for both one and two needles. Those made for one needle look similar to erasers that are attached to the ends of pencils. Those for two needles are wide and flat with two holes.

HOOK GAUGE

A hook gauge or *needle gauge* is a measuring device used by crocheters and knitters to test the sizes of particular crochet hooks and knitting needles. Hook gauges are usually made of plastic or aluminum and have sizing holes from 2mm to 11mm diameter.

A hook gauge also functions as a ruler to test the size of a test swatch of handmade fabric.

SIZING CONVENTIONS

A hook gauge is the most reliable way to determine the actual size of a tool. Even though most hooks and needles have metric sizing, tool dimensions vary slightly between different manufacturers. Non-metric sizing conventions also change over time. For instance, different sizing charts rate a 4.0mm hook as either a U.S. G/6 or an F/5.

IMPORTANCE OF ACCURATE SIZING

U.S. textile craft terminology employs the word *gauge* in two different ways: as *hook gauge* in reference to the sizing tool, or as a standalone term gauge to describe the number of stitches in a standard sized sample of work, usually 4" or 10cm. UK terminology refers to the latter as *tension*.

Craft patterns include a gauge or tension measurement and good practice among crafters is to achieve the correct gauge on a sample swatch before embarking on the full project.

As one crochet guidebook explains:

- Whenever you are following a crochet pattern you will notice that a gauge is included. This is an all-important guide to obtaining the proper size of the square or garment you are working on. If you don't get the proper gauge, the item will not fit properly and you may run out of yarn before finishing.

Several factors affect fabric gauge including hook or needle size, although sometimes a crafter may change tool size deliberately in order to achieve a certain gauge. Accurate needle and hook sizing reduces the number of variables that affect these decisions.

Additionally, if substitute tools are needed during the course of the project a hook gauge can determine whether a given tool is appropriate in advance of actual work.Row counter

(HAND KNITTING)

A row counter for hand knitting is a tally counter for counting rows or courses worked, for counting stitch pattern repetitions, or for counting increases or decreases of the number of stitches in consecutive rows. The first commercially produced one appeared on the market in the 1920s after the general public started regularly knitting from unfamiliar printed and complex patterns.

Design variations include on-needle barrel-shaped counters for straight-needle work, stitch-marker counters for knitting on double-pointed and circular needles, complex counters which attempted to assist with decreases, increases and lacework, stand-alone hand-held counters in imitation of the hand-tally, pendant counters worn round the neck and online software for iPhones.

EARLY TALLY METHODS IN EUROPE AND THE UNITED STATES

Until the early 19th century, in Europe and the United States, groups of localised professional hand knitters specialised in a few well-known patterns in which keeping a tally of rows was barely necessary, since the pattern and expected size of the work were known by heart.

From the early 19th century, when printed patterns were introduced for the recreational knitter, and until the early 20th century, woollen yarn had clear stitch definition so that rows were easily seen and counted, and a mental tally could be kept. However knitters could tally by moving objects from one pocket to another, or bracelets from one wrist to the other. For lace knitting and complicated increases, a knitter could write five-bar gate tally marks around the margins of a printed pattern.

ON-NEEDLE ROW COUNTERS

Flat-faced Type

The earliest on-needle row counters seem to have appeared in the UK between 1920 and 1939, when complicated printed patterns increased in popularity among the working population. In the 1920s a 1 inch (2.5 cm) grey-brown enamelled unit was manufactured with two rotary dials on the front showing tens and units, and slots on the back for sliding onto a knitting needle. The front was flat and shaped like a figure-8, but was heavier than the back so that the unit hung downward from the needle, making the numbered face difficult to read. A similar bakelite unit was manufactured by Abel Morrell for sale in the USA in the 1930s.

Around the 1980s to 1990s, *Woman's Weekly* magazine gave away a plastic on-needle knitting row counter of unusual design. It was presented as a self-assembly kit, in a small blue envelope with assembly instructions and a diagram on the back.

According to the diagram, the gadget consists of (a) two blue front and back plates, pierced to show the numbers; (b) an inner white soft plastic frame to hold the number discs and to which are attached the two loops to bend back and thread on the needle; (c) the two number discs. So the disks don't spin on an axis; they spin within circular spaces in the inner frame.

Rotary Barrel Type

In Britain in the 1950s when the baby boom years following World War II caused an increased requirement for hand-knitting, I.X. Products produced a brightly coloured and marketable plastic row counter with patent number 424432. This was a barrel-shaped counter which sat in stable fashion on the knitting needle by the aid of a central metal spring. The assemblage consisted of a double rotating inner barrel and an outer fixed and slotted skin which was pinned to the central spring.The two separately rotating halves of the inner barrel were printed with numbers, and the outer slotted skin rotated over the barrels, revealing a single ten and a single unit at one time. Precise engineering ensured that there was just enough ease of movement for the turning of the barrels, but just enough friction to prevent movement of the barrels while vigorous knitting continued.

However the introduction in the late 1950s of double knitting yarn and the consequent popular use of thicker needles tended to damage the springs which were intended for needles of a maximum thickness of 3.25mm. This design and most of the barrel-shaped on-needle counters which followed it had a white or cream central barrel and a decoratively coloured outer skin.In the 1950s and 1960s the UK knitting accessories manufacturer Millward produced the *Ro-Tally*, whose name was a pun on *rotary*. Its design was very similar to the I.X. Products version, it used the same patent number and the precision of engineering was improved, although the inner spring was still too small for 4 mm needles.

The Millward *Ro-Tally* was re-designed quite differently in the 1970s in various sizes, of softer plastic with a plastic spring which fitted large and small needles. These counters were not precisely engineered, so depended on polygonal barrels and outer skin to provide sufficient friction to control unnecessary movement while knitting.

The *Compact* rotary row counter was manufactured in England in the 1950s to 1960s under a pending patent application, although the general design was similar in principle to the early I.X. Products version. It had the same precise engineering for control of friction, and central metal spring suitable for needles up to 3.5 mm. The basic *Compact*, however, had a central knurled grip-band, which made the gadget easier to hold and to turn quickly and precisely. The earlier version of the late 1950s to early 1960s had red numerals, and the later 1960s version had black numerals.

The luxury version of the *Compact* was presented in a gift box, it was priced at a guinea and became a contemporary collector's item, tending to be little-used due to impracticality. The assemblage included a thin and delicate 6-inch - later 8-inch - tape measure in place of the grip band. On this luxury version the tape-measure assembly caused the centre of the gadget to rotate independently, making it difficult for adult fingers to grip the remaining part of the stable outer skin to turn the main assemblage for use as a row counter. Some of the counters in this range featured pearlised plastic outer skins in various pastel colours.

Around the 1960s to 1970s Woman's Weekly magazine gave away a soft plastic on-needle knitting row counter of unusual but simple design. Two numbered blue-and-white barrels rotate independently around a central white spindle, which has a number-pointer at one end and a loop to thread onto a needle at the other end. The friction (to prevent the barrels rolling freely) is maintained by the tightness of the number-pointer and shortness of the spindle.

In the 1980s the knitting accessories company Aero produced its own rotary on-needle counter which had a distinctive red inner barrel with thick knurled ends, and a grey outer skin. The red colour of the inner barrel meant that the numbers had to be printed in white, and were more easily visible than the small black printed numbers on earlier versions. Aero later produced a blue and white version of the same design, with the name Inox printed on it. In the 1990s Wrights Boyle and Prym began to produce a similar design with white inner barrels and coloured outer skin which continued to be sold in different sizes until at least 2010.

In the 21st century until at least 2010, the manufacturers Pony, Whitecroft and Wrights each made a new version of the 1970s octagonal counter for sale in Europe, the UK and the United States. At the same time the Japanese knitting accessories supplier Clover was exporting a version of the 1980s Aero counter, with dark inner barrels and white numbers, to the same customer base, although at first they supplied a pink version to the UK, and a blue version to the United States.

Jewellery Type

Commercial and home-crafted, beaded row counters appeared from 2000 until at least 2010. Some of these involved suspending jewellery from the needle and marking off beads as rows progressed.

Stitch-marker Row Counters

These tend to be on-needle barrel row counters which have been attached to stitch-marker rings.

They are intended to be used with circular or double-pointed knitting needles. In the 21st century Prym has marketed a dedicated stitch-marker row counter, labelling it the *Universal Rotally*.

COMPLEX ROW COUNTERS

In 1936 *The MP Handy Guide to Knitting and Crochet* was patented under number 813814 and manufactured in Britain as a pair of rectangular flat metal plates sandwiching a paper card marked with number lists, and sliders with markers to record current row number, increase, decrease and times.

However the metal slides did not work smoothly and could get stuck. The gadget incorporated a small ruler and needle gauge. There is an example of this counter in the Victoria and Albert Museum. The paper instructions pasted on the back carried the following legend:

- "Rows: When knitting and suddenly called away, mark off the number of rows reached.
- Increase/decrease: For casting on, or casting off stitches. Times: To mark off the pattern when doing fancy knitting, etc. Measure: Very handy when tape measure is mislaid.
- Needle gauge: For correct sizes from 6 to 12."

A paper card version of the *MP Handy Guide* was published with folded card sliders for the war effort under the auspices of the UK Ministry of Defence during World War II and was intended for use by people knitting gloves, socks and balaclavas for the troops. In the 1960s, The Aero Needles Group Plc of Redditch, England produced the *Knitters Companion*, a pocket version of the complex counter, consisting of a pierced bar of black plastic containing seven knurled, white, rotary, numbered discs whose numbers would appear through the piercings in the bar.

As in the *MP Handy Guide*, the numbers were labelled Increase, Decrease, Rows and Times. It was supplied with a soft plastic case which was intended to maintain memory of the count by preventing accidental movement of the discs. In 1972, UK magazine Woman's Weekly gave away its own card variation of this, with the numbers on dials with rotary paper pointers to record rows, increases or decreases, and times. Like the *MP Handy Guide* it incorporated a small inch-ruler and needle gauge, but now added a tension gauge measure and centimetre ruler.

In spring 2008 the UK magazine *Simply Knitting* no.39 gave away a green plastic version of the *MP Handy Guide*, without needle gauge or ruler and with a clarified version of the same instructions moulded onto the back:

- "Rows: move the markers when you have completed each row.
- Increase and decrease: keep count each time you need to increase or decrease stitches.
- Times: use the markers when you need to make more than one set of increases or decreases."

This version illustrated the difficulty of mass-producing such a gadget cheaply: the inner card is easily displaced during assembly, thereby misplacing the numbers, and some of the sliders do not move easily.

COUNTERS MADE OF CARD

These include the *World War II card counter*, and the 1972 *Woman's Weekly card counter*, both mentioned above. The wartime counter was a pierced card glued to a flat backing card; the front card contained strips printed with numerals and with slider-indicators. The numerals represented rows, increase or decrease, and times, as in previous complex counters; it also had a needle gauge and five-inch measure. The reverse contained instructions for use, and the legend: "a handy help for the war-time knitter". The wartime counter was a version of the 1930s *Bestway card counter.*

In the 1920s to 1930s, various card counters were produced in the UK, including the *Little Dorrit* counter, and the *schoolgirl* counter. From around 2006 Maik Remmers designed and manufactured in Germany a series of 3D card counters in the shape of skeins of wool, animals and abstract designs. These are folded and printed on back and front, with a pair of number discs sandwiched between the layers of card.

The discs can be turned to reveal numerals through two windows in the front layer of card. The outer surface of the card can be written-on with water-soluble ink, and the back surface of the counter contains a form where the knitter can fill in current row numbers. This is necessary as there is not enough friction to prevent the number discs from moving during storage.

VIYELLA NEEDLE GAUGE AND KNITTING COUNTER

The *Viyella needle gauge and knitting counter* was created between 1936 and 1940. It is sizeable and heavy at 1.5 inches (3.8 cm) high and 1.5 inches (3.8 cm) across, and is enamelled and chromed. The two flat, chromed ends of the drum are pierced to form a needle gauge. The ends rotate on the enamelled drum such that red or white enamelled numbers are revealed through piercings in the sides of both drum-ends. There is a black and a white version of this design and each has a transfer picture of two children in the style of Mabel Lucie Attwell. A third form, in black and without the picture, was possibly an austerity version produced during World War II.

PONY TOTAL KNITTING SYSTEM

The *Pony Total Knitting System* was manufactured by Pony in the 1990s, but ceased manufacture around 2008.

It consisted of a pair of short needles with flexible 13 inches (33 cm) cables attached to the ends, so that heavy knitted items did not have to be supported on long, straight needles. The ends of the cables were knobbed to prevent stitches falling off, and one cable had an on-needle-type barrel row-counter affixed to the end. However, unlike the standard on-needle row-counter, this counter was not removable, and there was a counter on one of every pair of needles of this design.

HANDHELD MECHANICAL AND ELECTRONIC ROW COUNTERS

The earliest of these, alongside the MP Handy Guide and Viyella counters above, was possibly the bakelite knitting pattern bookmark made in England in the 1930s, which incorporated a counter for 1-16 rows and a needle gauge. The large, red, mechanical *Kacha-Kacha* handheld counter of 2000-2010 by Clover is robustly made for adult hands and for older eyes which require large, clear numerals.

However it has no toddler-proof or drop-proof lock to maintain the count. Because it does not attach to a lanyard, it is more suited to the stationary workbasket than to travelling needlework.

However there is a similar design with a lock manufactured by KA in pink plastic for sale in the United States in the early 21st century, at least until 2010.

Pendant Knitting Counters

In the 21st century and at least until 2010, the Japanese company Clover was producing the bright green pendant *Mini Kacha-Kacha* which is designed like a hand-held mechanical tally counter, but is oriented upside-down on a neck pendant for easy reading by the user. It has a button to advance the numbers singly, access to the knurled number rings for adjustment of numbers, and a lock to prevent interference by the user dropping the gadget or by children playing with it.

On this type of gadget it has been possible for the manufacturer to make the numbers larger and therefore easier to read than on the on-needle counter, which needs to be small. In August 2007 the UK magazine *Simply Knitting* no.32 gave away a free pendant knitting counter in a form of a white sheep, and in June 2010 the same magazine no.67 gave away another one in the shape of a yellow flower. Unfortunately the numerals on the yellow flower version were oriented incorrectly, causing inconvenience for the wearer of the pendant. colored

Electronic Pendant Knitting Counters

The pendant *Electronic Row Counter* was produced in the early 21st century and was sold in the United States at least until 2010. Its LCD display counts up to 4 digits, whereas all foregoing knitting row counters were limited to 2 digits. It has add- and subtract-buttons which allow the user to count forwards and backwards in increments of one with an audible beep at each press, and the on-off button can be used to reset the count to zero. Because there is no memory facility, the counter must be left turned on if the count is to be maintained; however the gadget's instructions say that the battery will last for over seven months.

There is no lock so that the maintained count is vulnerable to playing children. The more complex version of this counter, the *Electronic Row Counter Plus*, has ceased manufacture.

STITCH HOLDER

Stitch holders are tools that are used by knitters to hold open stitches when not being used by the needles. They can be used when finishing a side of an item, such as a sweater, and preparing for the kitchener stitch. Then can be used on stitches separated from the needle when splitting the piece. For example, when shaping the neckhole of a sweater, the stitch holder is placed on the group of stitches used to make the neckhole, and the needles are used to continue to make the stitches on either side. Stitch holders are also used at the ends of sleeves of a sweater before they are attached to the body of the sweater.

2

Yarn Styles, Brands and Patterns

KNITTING

Knitting is a method by which thread or yarn is used to create a cloth. Knitted fabric consists of a number of consecutive rows of loops, called stitches. As each row progresses, a new loop is pulled through an existing loop. The active stitches are held on a needle until another loop can be passed through them. This process eventually results in a fabric, often used for blankets or garments.

Knitting may be done by hand or by machine. There exist numerous styles and methods of hand knitting. Different types of yarns and needles may be used to achieve a plethora of knitted materials; these tools give the final piece a different colour, texture, weight, and/or integrity. Other factors that affect the end result include the needle's shape, thickness and malleability, as well as the yarn's fibre type, texture and twist.

STRUCTURE

Courses and Wales

Like weaving, knitting is a technique for producing a two-dimensional fabric made from a one-dimensional yarn or thread. In weaving, threads are always straight, running parallel either lengthwise (warp threads) or crosswise (weft threads). By contrast, the yarn in knitted fabrics follows a meandering path (a *course*), forming symmetric loops (also called bights) symmetrically above and below the mean path of the yarn.

These meandering loops can be easily stretched in different directions giving knit fabrics much more elasticity than woven fabrics. Depending on the yarn and knitting pattern, knitted garments can stretch as much as 500 per cent. For this reason, knitting was initially developed for garments that must be elastic or stretch in response to the wearer's motions, such as socks and hosiery. For comparison, woven garments stretch mainly along one or other of a related pair of directions that lie roughly diagonally between the warp and the weft, while contracting in the other direction of the pair (stretching and contracting with the *bias*), and are not very elastic, unless they are woven from stretchable material such as spandex.

Knitted garments are often more form-fitting than woven garments, since their elasticity allows them to contour to the body's outline more closely; by contrast, curvature is introduced into most woven garments only with sewn darts, flares, gussets and gores, the seams of which lower the elasticity of the woven fabric still further. Extra curvature can be introduced into knitted garments without seams, as in the heel of a sock; the effect of darts, flares, etc. can be obtained with short rows or by increasing or decreasing the number of stitches. Thread used in weaving is usually much finer than the yarn used in knitting, which can give the knitted fabric more bulk and less drape than a woven fabric.

If they are not secured, the loops of a knitted course will come undone when their yarn is pulled; this is known as *ripping out*, *unravelling* knitting, or humorously, *frogging* (because you 'rip it', this sounds like a frog croaking: 'rib-bit'). To secure a stitch, at least one new loop is passed through it. Although the new stitch is itself unsecured ("active" or "live"), it secures the stitch(es) suspended from it. A sequence of stitches in which each stitch is suspended from the next is called a *wale*.

To secure the initial stitches of a knitted fabric, a method for casting on is used; to secure the final stitches in a wale, one uses a method of binding/casting off. During knitting, the active stitches are secured mechanically, either from individual hooks (in knitting machines) or from a knitting needle or frame in hand-knitting.

Weft and Warp Knitting

There are two major varieties of knitting: weft knitting and warp knitting. In the more common *weft knitting*, the wales are perpendicular to the course of the yarn. In warp knitting, the wales and courses run roughly parallel. In weft knitting, the entire fabric may be produced from a single yarn, by adding stitches to each wale in turn, moving across the fabric as in a raster scan. By contrast, in warp knitting, one yarn is required for every wale. Since a typical piece of knitted fabric may have hundreds of wales, warp knitting is typically done by machine, whereas weft knitting is done by both hand and machine. Warp-knitted fabrics such as tricot and milanese are resistant to runs, and are commonly used in lingerie.

Weft-knit fabrics may also be knit with multiple yarns, usually to produce interesting colour patterns. The two most common approaches are intarsia and stranded colourwork. In intarsia, the yarns are used in well-segregated regions, *e.g.*, a red apple on a field of green; in that case, the yarns are kept on separate spools and only one is knitted at any time. In the more complex stranded approach, two or more yarns alternate repeatedly within one row and all the yarns must be carried along the row. Double knitting can produce two separate knitted fabrics simultaneously (*e.g.*, two socks). However, the two fabrics are usually integrated into one, giving it great warmth and excellent drape.

Knit and Purl Stitches

In securing the previous stitch in a wale, the next stitch can pass through the previous loop from either below or above. If the former, the stitch is denoted as a *knit stitch* or a *plain stitch*; if the latter, as a *purl stitch*. The two stitches are related in that a knit stitch seen from one side of the fabric appears as a purl stitch on the other side.

The two types of stitches have a different visual effect; the knit stitches look like "V"'s stacked vertically, whereas the purl stitches look like a wavy horizontal line across the fabric. Patterns and pictures can be created in knitted fabrics by using knit and purl stitches as "pixels"; however, such pixels are usually rectangular, rather than square, depending on the gauge/tension of the knitting. Individual stitches, or rows of stitches, may be made taller by drawing more yarn into the new loop (an elongated stitch), which is the basis for uneven knitting: a row of tall stitches may alternate with one or more rows of short stitches for an interesting visual effect. Short and tall stitches may also alternate within a row, forming a fish-like oval pattern.

In the simplest knitted fabrics, all the stitches are knit or purl; this is known as a garter stitch. Alternating rows of knit stitches and purl stitches produce what is known as a stockinette pattern/stocking stitch. Vertical stripes (ribbing) are possible by having alternating wales of knit and purl stitches. For example, a common choice is 2x2 ribbing, in which two wales of knit stitches are followed by two wales of purl stitches, etc.

Horizontal striping (welting) is also possible, by alternating *rows* of knit and purl stitches. Checkerboard patterns (basketweave) are also possible, the smallest of which is known: the stitches alternate between knit and purl in every wale and along every row.

Fabrics in which the number of knit and purl stitches are not the same, such as stockinette/stocking stitch, have a tendency to curl; by contrast, those in which knit and purl stitches are arranged symmetrically (such as ribbing, garter stitch or seed/moss stitch) tend to lie flat and drape well. Wales of purl stitches have a tendency to recede, whereas those of knit stitches tend to come forward. Thus, the purl wales in ribbing tend to be invisible, since the neighbouring knit wales come forward. Conversely, rows of purl stitches tend to form an embossed ridge relative to a row of knit stitches. This is the basis of shadow knitting, in which the appearance of a knitted fabric changes when viewed from different directions.

Typically, a new stitch is passed through a single unsecured ("active") loop, thus lengthening that wale by one stitch. However, this need not be so; the new loop may be passed through an already secured stitch lower down on the fabric, or even between secured stitches (a dip stitch). Depending on the distance between where the loop is drawn through the fabric and where it is knitted, dip stitches can produce a subtle stippling or long lines across the surface of the fabric, *e.g.*, the lower leaves of a flower. The new loop may also be passed between two stitches in the *present* row, thus clustering the intervening stitches; this approach is often used to produce a smocking effect in the fabric.

The new loop may also be passed through *two or more* previous stitches, producing a decrease and merging wales together. The merged stitches need not be from the same row; for example, a tuck can be formed by knitting stitches together from two different rows, producing a raised horizontal welt on the fabric.

Not every stitch in a row need be knitted; some may be left "as is" and knitted on a subsequent row. This is known as slip-stitch knitting. The slipped stitches are naturally longer than the knitted ones. For example, a stitch slipped for one row before knitting would be roughly twice as tall as its knitted counterparts. This can produce interesting visual effects, although the resulting fabric is more rigid because the slipped stitch "pulls" on its neighbours and is less deformable. Slip-stitch knitting plays an important role in *mosaic knitting*, an important technique in hand-knitting patterned fabrics; mosaic-knit fabrics tend to be stiffer than patterned fabrics produced by other methods such as Fair-Isle knitting.

In some cases, a stitch may be deliberately left unsecured by a new stitch and its wale allowed to disassemble. This is known as drop-stitch knitting, and produces a vertical ladder of see-through holes in the fabric, corresponding to where the wale had been.

Right- and Left-plaited Stitches

Both knit and purl stitches may be twisted: usually once if at all, but sometimes twice and (very rarely) thrice. When seen from above, the twist can be clockwise (right yarn over left) or counterclockwise (left yarn over right); these are denoted as right- and left-plaited stitches, respectively. Hand-knitters generally produce right-plaited stitches by knitting or purling through the back loops, i.e., passing the needle through the initial stitch in an unusual way, but wrapping the yarn as usual. By contrast, the left-plaited stitch is generally formed by hand-knitters by wrapping the yarn in the opposite way, rather than by any change in the needle. Although they are mirror images in form, right- and left-plaited stitches are functionally equivalent. Both types of plaited stitches give a subtle but interesting visual texture, and tend to draw the fabric inwards, making it stiffer. Plaited stitches are a common method for knitting jewelry from fine metal wire.

Edges and Joins Between Fabrics

The initial and final edges of a knitted fabric are known as the *cast-on* and *bound/cast-off* edges. The side edges are known as the *selvages*; the word derives from "self-edges", meaning that the stitches do not need to be secured by anything else. Many types of selvages have been developed, with different elastic and ornamental properties.

Vertical and horizontal edges can be introduced within a knitted fabric, *e.g.*, for button holes, by binding/casting off and re-casting on again (horizontal) or by knitting the fabrics on either side of a vertical edge separately.

Two knitted fabrics can be joined by embroidery-based grafting methods, most commonly the Kitchener stitch. New wales can be begun from any of the edges of a knitted fabric; this is known as picking up stitches and is the basis for entrelac, in which the wales run perpendicular to one another in a checkerboard pattern.

Cables, Increases, and Lace

Ordinarily, stitches are knitted in the same order in every row, and the wales of the fabric run parallel and vertically along the fabric. However, this need not be so, since the order in which stitches are knitted may be permuted so that wales cross over one another, forming a cable pattern. Cables patterns tend to draw the fabric together, making it denser and less elastic; Aran sweaters are a common form of knitted cabling.

Arbitrarily complex braid patterns can be done in cable knitting, with the proviso that the wales must move ever upwards; it is generally impossible for a wale to move up and then down the fabric. Knitters have developed methods for giving the illusion of a circular wale, such as appear in Celtic knots, but these are inexact approximations. However, such circular wales are possible using Swiss darning, a form of embroidery, or by knitting a tube separately and attaching it to the knitted fabric.

A wale can split into two or more wales using increases, most commonly involving a yarn over. Depending on how the increase is done, there is often a hole in the fabric at the point of the increase.

This is used to great effect in lace knitting, which consists of making patterns and pictures using such holes, rather than with the stitches themselves. The large and many holes in lacy knitting makes it extremely elastic; for example, some Shetland "wedding-ring" shawls are so fine that they may be drawn through a wedding ring.

By combining increases and decreases, it is possible to make the direction of a wale slant away from vertical, even in weft knitting. This is the basis for bias knitting, and can be used for visual effect, similar to the direction of a brush-stroke in oil painting.

Ornamentations and Additions

Various point-like ornaments may be added to knitting for their look or to improve the wear of the fabric. Examples include various types of bobbles, sequins and beads. Long loops can also be drawn out and secured, forming a "shaggy" texture to the fabric; this is known as loop knitting. Additional patterns can be made on the surface of the knitted fabric using embroidery; if the embroidery resembles knitting, it is often called Swiss darning. Various closures for the garments, such as frogs and buttons can be added; usually buttonholes are knitted into the garment, rather than cut.

Ornamental pieces may also be knitted separately and then attached using applique. For example, differently coloured leaves and petals of a flower could be knit separately and attached to form the final picture. Separately knitted tubes can be applied to a knitted fabric to form complex Celtic knots and other patterns that would be difficult to knit.

Unknitted yarns may be worked into knitted fabrics for warmth, as is done in tufting and "weaving" (also known as "couching").

TYPES

Flat Knitting Versus Circular Knitting

Flat knitting, in its most basic form, is used to make flat pieces of cloth. It is done with two straight knitting needles and is worked in rows, horizontal lines of stitches. Flat knitting is usually used to knit flat pieces like scarves, blankets, afghans, and the backs and fronts of sweaters and pullovers.

Circular knitting (also called "knitting in the round") is a form of knitting that can be used to create a seamless tube. Knitting is worked in rounds (the equivalent of rows in flat knitting). Originally, circular knitting was done using a set of four or five double-pointed knitting needles.

Circular needles were later invented making this type of knitting easier. A circular needle resembles two short knitting needles connected by a cable of varying length between them. A circular knitting needle with a long cable can be used in place of straight needles to create larger flat-knitted pieces of fabric. Both types of circular knitting are used in creating pieces that are circular or tube-shaped, such as hats, socks, mittens, and sleeves.

Arm Knitting

This kind of knitting uses one's arms in place of knitting needles. Stitches end up much larger and the resulting fabric has a lot of drape.

- "The resulting scarves and blankets feature rows of loops that are 2.5" to 4" wide, depending on the diameter of a given knitter's arm. Owing to the large, holey rows, an entire project can be completed in less than 30 minutes—a fraction of the time needed to whip up a scarf using knitting needles.

Finger Knitting

This type of knitting uses one's fingers to construct knitted fabric that looks like i-cord with larger stitches. Finger knitting is useful for making headbands, belts, or other cord-like objects.

Felting

Felting is the hand-knitters' term for fulling, a technique for joining knitted or woven wool fibres (only animal yarn fibres can be felted. Acrylic or any fibre that can be machine washed cannot be felted).

The finished product is put in a hot wash and the yarn fibres become agitated until it starts to shrink. While the knitted project is in the washing machine, it is important to check the progress. Depending on the size of the project, some may felt faster than others. The end result typically has reduced dimensions. Bags, mittens, vests, socks, slippers and hats are just a few items that can be felted.

Needle Felting

Needle felting is a technique used to add decoration to a knitted or felted piece, where raw roving is applied using a very sharp barbed felting needle by repeatedly piercing the roving and background together. Once washed in hot water, the appliqued decoration is fused with the background. Felted knitting can be cut with scissors without concern about fraying. This is also a good method to use when attempting to give structure to what can be frail knitted fabrics.

HISTORY AND CULTURE

The word is derived from *knot* and ultimately from the Old English *cnyttan*, to knot.

Nålebinding (Danish: literally "binding with a needle" or "needle-binding") is a fabric creation technique predating both knitting and crochet. That said, one of the earliest known examples of true knitting was cotton socks with stranded knit colour patterns found in Egypt from the end of the first millennium

AD. Originally a male-only occupation, the first knitting trade guild was started in Paris in 1527. With the invention of the knitting machine, knitting "by hand" became a craft used by country people with easy access to fiber. Similar to quilting, spinning, and needlepoint, knitting became a leisure activity for the wealthy.

Hand-knitting has gone into and out of fashion many times in the last two centuries, and at the turn of the 21st century it is enjoying a revival. According to the industry group Craft Yarn Council of America, the number of women knitters in the United States age 25–35 increased 150 per cent in the two years between 2002 and 2004. The latest incarnation is less about the "make-do and mend" attitude of the 1940s and early 50s and more about making a statement about individuality as well as developing an innate sense of community.

During the 1940s, English knitting rose in popularity while Continental knitting fell. This is due to the fact that continental knitting originated within Germany and was spread by immigrants. During World War II, continental knitting fell out of style due to its relationship with Germany. It wasn't until Elizabeth Zimmermann publicized continental knitting in the 1980s that it again was popularized.

Additionally, many contemporary knitters have an interest in blogging about their knitting, patterns, and techniques, or joining a virtual community focused on knitting, such as the extremely popular Ravelry. There are also a number of popular knitting podcasts, and various other knitting web sites. Contemporary knitting groups may be referred to in the U.S. as a "Stitch 'N Bitch" where a group of knitters get together to work on projects, discuss patterns, troubleshoot their work and just socialize. In the UK, the term has been "knitting circle" since the early 20th century.

PROPERTIES OF FABRICS

The topology of a knitted fabric is relatively complex. Unlike woven fabrics, where strands usually run straight horizontally and vertically, yarn that has been knitted follows a looped path along its row, as with the red strand in the diagram at left, in which the loops of one row have all been pulled through the loops of the row below it.

Because there is no single straight line of yarn anywhere in the pattern, a knitted piece of fabric can stretch in all directions. This elasticity is all but unavailable in woven fabrics which only stretch along the bias. Many modern stretchy garments, even as they rely on elastic synthetic materials for some stretch, also achieve at least some of their stretch through knitted patterns.

The basic knitted fabric (as in the diagram, and usually called a *stocking* or *stockinette* pattern) has a definite "right side" and "wrong side". On the right side, the visible portions of the loops are the verticals connecting two rows which are arranged in a grid of *V* shapes. On the wrong side, the ends of the loops are visible, both the tops and bottoms, creating a much more bumpy texture sometimes called *reverse stockinette*. (Despite being the "wrong side," reverse stockinette is frequently used as a pattern in its own right.) Because the yarn holding rows together is all on the front, and the yarn holding side-by-side stitches together is all on the back, stockinette fabric has a strong tendency to curl towards the front on the top and bottom, and towards the back on the left and right side.

Stitches can be worked from either side, and various patterns are created by mixing regular knit stitches with the "wrong side" stitches, known as purl stitches, either in columns (ribbing), rows (garter, welting), or more complex patterns. Each fabric has different properties: a garter stitch has much more vertical stretch, while ribbing stretches much more horizontally. Because of their front-back symmetry, these two fabrics have little curl, making them popular as edging, even when their stretch properties are not desired.

Different combinations of knit and purl stitches, along with more advanced techniques, generate fabrics of considerably variable consistency, from gauzy to very dense, from highly stretchy to relatively stiff, from flat to tightly curled, and so on.

Texture

The most common texture for a knitted garment is that generated by the flat stockinette stitch, though very small, in machine-made stockings and T-shirts—which is worked in the round as nothing but knit stitches, and worked flat as alternating rows of knit and purl. Other simple textures can be made with nothing but knit and purl stitches, including garter stitch, ribbing, and moss and seed stitches. Adding a "slip stitch" (where a loop is passed from one needle to the other) allows for a wide range of textures, including heel and linen stitches as well as a number of more complicated patterns.

Some more advanced knitting techniques create a surprising variety of complex textures. Combining certain increases, which can create small eyelet holes in the resulting fabric, with assorted decreases is key to creating knitted lace, a very open fabric resembling lace. Changing the order of stitches from one row to the next, usually with the help of a cable needle or stitch holder, is key to cable knitting, producing an endless variety of cables, honeycombs, ropes, and Aran sweater patterning. Entrelac forms a rich checkerboard texture by knitting small squares, picking up their side edges, and knitting more squares to continue the piece.

The appearance of a garment is also affected by the *weight* of the yarn, which describes the thickness of the spun fibre. The thicker the yarn, the more visible and apparent stitches will be; the thinner the yarn, the finer the texture.

Colour

Plenty of finished knitting projects never use more than a single colour of yarn, but there are many ways to work in multiple colours. Some yarns are dyed to be either *variegated* (changing colour every few stitches in a random fashion) or *self-striping* (changing every few rows). More complicated techniques permit large fields of colour (intarsia, for example), busy small-scale patterns of colour (such as Fair Isle), or both (double knitting and slip-stitch colour, for example).

Yarn with multiple shades of the same hue are called *ombre*, while a yarn with multiple hues may be known as a given *colourway* — a green, red and yellow yarn might be dubbed the "Parrot Colourway" by its manufacturer, for example. *Heathered* yarns contain small amounts of fibre of different colours, while *tweed* yarns may have greater amounts of different coloured fibres.

PROCESS

There are many hundreds of different knitting stitches used by knitters. A piece of knitting begins with the process of *casting on*, which involves the initial creation of the stitches on the needle. Different methods of casting on are used for different effects: one may be stretchy enough for lace, while another provides a decorative edging. *Provisional* cast-ons are used when the knitting will continue in both directions from the cast-on.

There are various methods employed to cast on, such as the "thumb method" (also known as "slingshot" or "long-tail" cast-ons), where the stitches are created by a series of loops that will, when knitted, give a very loose edge ideal for "picking up stitches" and knitting a border; the "double needle method" (also known as "knit-on" or "cable cast-on"), whereby each loop placed on the needle is then "knitted on," which produces a firmer edge ideal on its own as a border; and many more. The number of active stitches remains the same as when cast on unless stitches are added (an increase) or removed (a decrease).

Most Western-style knitters follow either the English style (in which the yarn is held in the right hand) or the Continental style (in which the yarn is held in the left hand).

There are also different ways to insert the needle into the stitch. Knitting through the front of a stitch is called Western knitting. Going through the back of a stitch is called Eastern knitting. A third method, called combination knitting, goes through the front of a knit stitch and the back of a purl stitch.

Once the knitted piece is finished, the remaining live stitches are "cast off". Casting (or "binding") off loops the stitches across each other so they can be removed from the needle without unravelling the item. Although the mechanics are different from casting on, there is a similar variety of methods.

In knitting certain articles of clothing, especially larger ones like sweaters, the final knitted garment will be made of several knitted pieces, with individual sections of the garment knit separately and then sewn together. Seamless knitting, where a whole garment is knit as a single piece, is also possible. Elizabeth Zimmermann is probably the best-known proponent of seamless or circular knitting techniques. Smaller items, such as socks and hats, are usually knit in one piece on double-pointed needles or circular needles. (See *Circular knitting*.)

MEGA KNITTING

Mega knitting is a term recently coined and relates to the use of knitting needles greater than or equal to half an inch in diameter.

Mega knitting uses the same stitches and techniques as conventional knitting, except that hooks are carved into the ends of the needles. The hooked needles greatly enhance control of the work, catching the stitches and preventing them from slipping off.

It was the development of the knitting machine that introduced hooked needles and enabled faultless, automated knitting. Most knitters probably aren't even aware of the many processes that their fingers perform in the making of a single stitch. However, large gauge needles emphasise those actions and knitting becomes increasingly more awkward when the needle diameter is greater than the width of the knitters finger. On a one inch diameter (size 50) needle for instance, the shaft begins to taper one and three quarter inches from the tip. This means that the stitches are spread much further apart on mega knitting needles, making them more difficult to control.

The hook catches the loop of yarn as each stitch is knitted, meaning that wrists and fingers don't have to work so hard and there is less chance of stitches slipping off the needle. The position of the hook is most important. Turn the left (non-working) hook to face away at all times; turn the right (working) hook towards you up whilst knitting (plain stitch) and away whilst purling. Mega knitting produces a chunky, bulky fabric or an open lacy weave, depending on the weight and type of yarn used.

MATERIALS

Yarn

Yarn for hand-knitting is usually sold as balls or skeins (hanks), although it may also be wound on spools or cones. Skeins and balls are generally sold with a *yarn-band*, a label that describes the yarn's weight, length, dye lot, fiber content, washing instructions, suggested needle size, likely gauge/tension, etc. It is common practice to save the yarn band for future reference, especially if additional skeins must be purchased. Knitters generally ensure that the yarn for a project comes from a single dye lot. The dye lot specifies a group of skeins that were dyed together and thus have precisely the same colour; skeins from different dye-lots, even if very similar in colour, are usually slightly different and may produce a visible stripe when knitted together. If a knitter buys insufficient yarn of a single dye lot to complete a project, additional skeins of the same dye lot can sometimes be obtained from other yarn stores or online. Otherwise, knitters can alternate skeins every few rows to help the dye lots blend together easier.

The thickness or weight of the yarn is a significant factor in determining the gauge/tension, *i.e.*, how many stitches and rows are required to cover a given area for a given stitch pattern. Thicker yarns generally require thicker knitting needles, whereas thinner yarns may be knit with thick or thin needles. Hence, thicker yarns generally require fewer stitches, and therefore less time, to knit up a given garment.

Patterns and motifs are coarser with thicker yarns; thicker yarns produce bold visual effects, whereas thinner yarns are best for refined patterns. Yarns are grouped by thickness into six categories: superfine, fine, light, medium, bulky and superbulky; quantitatively, thickness is measured by the number of wraps per inch (WPI). In the British Commonwealth (outside North America) yarns are measured as 1ply, 2ply, 3ply, 4ply, 5ply, 8ply (or double knit),10ply and 12ply (triple knit). The related *weight per unit length* is usually measured in tex or denier.

Before knitting, the knitter will typically transform a hank/skein into a ball where the yarn emerges from the center of the ball; this making the knitting easier by preventing the yarn from becoming easily tangled. This transformation may be done by hand, or with a device known as a ballwinder. When knitting, some knitters enclose their balls in jars to keep them clean and untangled with other yarns; the free yarn passes through a small hole in the jar-lid.

A yarn's usefulness for a knitting project is judged by several factors, such as its *loft* (its ability to trap air), its *resilience* (elasticity under tension), its washability and colourfastness, its *hand* (its feel, particularly softness vs. scratchiness), its durability against abrasion, its resistance to pilling, its *hairiness* (fuzziness), its tendency to twist or untwist, its overall weight and drape, its blocking and felting qualities, its comfort (breathability, moisture absorption, wicking properties) and of course its look, which includes its colour, sheen, smoothness and ornamental features.

Other factors include allergenicity; speed of drying; resistance to chemicals, moths, and mildew; melting point and flammability; retention of static electricity; and the propensity to become stained and to accept dyes. Different factors may be more significant than others for different knitting projects, so there is no one "best" yarn. The resilience and propensity to (un)twist are general properties that affect the ease of hand-knitting. More resilient yarns are more forgiving of irregularities in tension; highly twisted yarns are sometimes difficult to knit, whereas untwisting yarns can lead to split stitches, in which not all the yarn is knitted into a stitch.

A key factor in knitting is *stitch definition*, corresponding to how well complicated stitch patterns can be seen when made from a given yarn. Smooth, highly spun yarns are best for showing off stitch patterns; at the other extreme, very fuzzy yarns or eyelash yarns have poor stitch definition, and any complicated stitch pattern would be invisible.

Although knitting may be done with ribbons, metal wire or more exotic filaments, most yarns are made by spinning fibers. In spinning, the fibers are twisted so that the yarn resists breaking under tension; the twisting may be done in either direction, resulting in a Z-twist or S-twist yarn. If the fibers are first aligned by combing them, the yarn is smoother and called a *worsted*; by contrast, if the fibers are carded but not combed, the yarn is fuzzier and called *woolen-spun*.

The fibers making up a yarn may be continuous *filament* fibers such as silk and many synthetics, or they may be *staples* (fibers of an average length, typically a few inches); naturally filament fibers are sometimes cut up into staples before spinning. The strength of the spun yarn against breaking is determined by the amount of twist, the length of the fibers and the thickness of the yarn. In general, yarns become stronger with more twist (also called *worst*), longer fibers and thicker yarns (more fibers); for example, thinner yarns require more twist than do thicker yarns to resist breaking under tension. The thickness of the yarn may vary along its length; a *slub* is a much thicker section in which a mass of fibers is incorporated into the yarn.

The spun fibers are generally divided into animal fibers, plant and synthetic fibers. These fiber types are chemically different, corresponding to proteins, carbohydrates and synthetic polymers, respectively.

Animal fibers include silk, but generally are long hairs of animals such as sheep (wool), goat (angora, or cashmere goat), rabbit (angora), llama, alpaca, dog, cat, camel, yak, and muskox (qiviut). Plants used for fibers include cotton, flax (for linen), bamboo, ramie, hemp, jute, nettle, raffia, yucca, coconut husk, banana fiber, soy and corn. Rayon and acetate fibers are also produced from cellulose mainly derived from trees.

Common synthetic fibers include acrylics, polyesters such as dacron and ingeo, nylon and other polyamides, and olefins such as polypropylene. Of these types, wool is generally favored for knitting, chiefly owing to its superior elasticity, warmth and (sometimes) felting; however, wool is generally less convenient to clean and some people are allergic to it. It is also common to blend different fibers in the yarn, *e.g.*, 85 per cent alpaca and 15 per cent silk. Even within a type of fiber, there can be great variety in the length and thickness of the fibers; for example, Merino wool and Egyptian cotton are favored because they produce exceptionally long, thin (fine) fibers for their type.

A single spun yarn may be knitted as is, or braided or plied with another. In plying, two or more yarns are spun together, almost always in the opposite sense from which they were spun individually; for example, two Z-twist yarns are usually plied with an S-twist.

The opposing twist relieves some of the yarns' tendency to curl up and produces a thicker, *balanced* yarn. Plied yarns may themselves be plied together, producing *cabled yarns* or *multi-stranded yarns*. Sometimes, the yarns being plied are fed at different rates, so that one yarn loops around the other, as in bouclé. The single yarns may be dyed separately before plying, or afterwards to give the yarn a uniform look.

The dyeing of yarns is a complex art that has a long history. However, yarns need not be dyed. They may be dyed just one colour, or a great variety of colours. Dyeing may be done industrially, by hand or even hand-painted onto the yarn. A great variety of synthetic dyes have been developed since the synthesis of indigo dye in the mid-19th century; however, natural dyes are also possible, although they are generally less brilliant. The colour-scheme of a yarn is sometimes called its colourway. Variegated yarns can produce interesting visual effects, such as diagonal stripes; conversely, a variegated yarn may frustrate an otherwise good knitting pattern by producing distasteful colour combination.

Glass/Wax

Knitted Glass combines knitting, lost-wax casting, mold-making, and kiln-casting. The process involves

1. *knitting* with wax strands,
2. surrounding the knitted wax piece with a heat-tolerant refractory material,
3. removing the wax by melting it out, thus creating a mold;
4. placing the mold in a kiln where lead crystal glass melts into the mold;
5. after the mold cools, the mold material is removed to reveal the knitted glass piece.

TOOLS

The process of knitting has three basic tasks:

1. the active (unsecured) stitches must be held so they don't drop
2. these stitches must be released sometime after they are secured

3. new bights of yarn must be passed through the fabric, usually through active stitches, thus securing them.

In very simple cases, knitting can be done without tools, using only the fingers to do these tasks; however, knitting is usually carried out using tools such as knitting needles, knitting machines or rigid frames. Depending on their size and shape, the rigid frames are called knitting boards, knitting rings (also called knitting looms) or knitting spools (also known as knitting knobbies, knitting nancies, or corkers). There is also a technique of knitting with a crochet hook that has a cord attached to the end, to hold the stitches while they're being worked. This technique is called knooking. Other tools are used to prepare yarn for knitting, to measure and design knitted garments, or to make knitting easier or more comfortable.

Needles

There are three basic types of knitting needles (also called "knitting pins"). The first and most common type consists of two slender, straight sticks tapered to a point at one end, and with a knob at the other end to prevent stitches from slipping off. Such needles are usually 10–16 inches (250–410 mm) long but, due to the compressibility of knitted fabrics, may be used to knit pieces significantly wider.

The most important property of needles is their diameter, which ranges from below 2 to 25 mm (roughly 1 inch). The diameter affects the size of stitches, which affects the gauge/tension of the knitting and the elasticity of the fabric. Thus, a simple way to change gauge/tension is to use different needles, which is the basis of uneven knitting.

Although the diameter of the knitting needle is often measured in millimeters, there are several measurement systems, particularly those specific to the United States, the United Kingdom and Japan; a conversion table is given at knitting needle. Such knitting needles may be made out of any materials, but the most common materials are metals, wood, bamboo, and plastic. Different materials have different frictions and grip the yarn differently; slick needles such as metallic needles are useful for swift knitting, whereas rougher needles such as bamboo offer more friction and are therefore less prone to dropping stitches.

The knitting of new stitches occurs only at the tapered ends. Needles with lighted tips have been sold to allow knitters to knit in the dark.The second type of knitting needles are straight, double-pointed knitting needles (also called "DPNs").

Double-pointed needles are tapered at both ends, which allows them to be knit from either end. DPNs are typically used for circular knitting, especially smaller tube-shaped pieces such as sleeves, collars, and socks; usually one needle is active while the others hold the remaining active stitches. DPNs are somewhat shorter (typically 7 inches) and are usually sold in sets of four or five.

Cable needles are a special case of DPNs, although they are usually not straight, but dimpled in the middle. Often, they have the form of a hook. When cabling a knitted piece, a hook is easier to grab and hold the yarn. Cable needles are typically very short (a few inches), and are used to hold stitches temporarily while others are being knitted. Cable patterns are made by permuting the order of stitches; although one or two stitches may be held by hand or knit out of order, cables of three or more generally require a cable needle.

The third needle type consists of circular needles, which are long, flexible double-pointed needles. The two tapered ends (typically 5 inches (130 mm) long) are rigid and straight, allowing for easy knitting; however, the two ends are connected by a flexible strand (usually nylon) that allows the two ends to be brought together. Circular needles are typically 24-60 inches long, and are usually used singly or in pairs; again, the width of the knitted piece may be significantly longer than the length of the circular needle.

A developing trend in the knitting world is interchangeable needles. These kits consist of pairs of needles with usually nylon cables or cords. The cables/cords are screwed into the needles, allowing the knitter to have both flexible straight needles or circular needles. This also allows the knitter to change the diameter and length of the needles as needed.

The ability to work from either end of one needle is convenient in several types of knitting, such as slip-stitch versions of double knitting. Circular needles may be used for flat or circular knitting.

Cable needles are a specific design, and are used to create the twisting motif of a knitted cable. They are made in different sizes, which produces cables of different widths. When in use, the cable needle is used at the same time as two regular needles.

It functions by holding together the stitches creating the cable as the other needles create the rest of the stitches for the knitted piece. At specific points indicated by the pattern, the cable needle is moved, the stitches on it are worked by the other needles, then the cable needle is turned around to a different position to create the cable twist.

Ancillary Tools

Various tools have been developed to make hand-knitting easier. Tools for measuring needle diameter and yarn properties have been discussed above, as well as the yarn swift, ballwinder and "yarntainers". Crochet hooks and a darning needle are often useful in binding/casting off or in joining two knitted pieces edge-to-edge. The darning needle is used in duplicate stitch (also known as Swiss darning), while the crochet hook is also essential for repairing dropped stitches and some specialty stitches such as tufting. Other tools are used to prepare specific ornaments include the pompom tree for making pompoms conveniently.

For large or complex patterns, it is sometimes difficult to keep track of which stitch should be knit in a particular way; therefore, several tools have been developed to identify the number of a particular row or stitch, including circular stitch markers, hanging markers, extra yarn and row counters. A second potential difficulty is that the knitted piece will slide off the tapered end of the needles when unattended; this is prevented by "point protectors" that cap the tapered ends.

Another problem is that too much knitting may lead to hand and wrist troubles; for this, special stress-relieving gloves are available. In traditional Shetland knitting a special belt is often used to support the end of one needle allowing the knitting greater speed. Finally, there are sundry bags and containers for holding knitting, yarns and needles.

COMMERCIAL APPLICATIONS

Industrially, metal wire is also knitted into a metal fabric for a wide range of uses including the filter material in cafetieres, catalytic converters for cars and many other uses. These fabrics are usually manufactured on circular knitting machines that would be recognised by conventional knitters as sock machines. Many fashion designers make heavy use of knitting in their fashion collections. Gordana Gelhausen, who appeared in season six of the television show *Project Runway*, is primarily a knit designer. Other designers and labels that make heavy use of knitting include Michael Kors, Fendi, and Marc Jacobs. For individual hobbyists, web sites such as Etsy,Big Cartel and Ravelry have made it easy to sell patterns on a small scale, in a way similar to eBay.

GRAFFITI

In the last decade, a practice called knitting graffiti, guerilla knitting, or yarn bombing— the use of knitted or crocheted cloth to modify and beautify one's (usually outdoor) surroundings—emerged in the U.S. and spread worldwide. Magda Sayeg is credited with starting the movement in the US and Knit the City are a prominent group of graffiti knitters in the United Kingdom. Yarn bombers sometimes target existing pieces of graffiti for beautification. For instance, Dave Cole is a contemporary sculpture artist who practiced knitting as graffiti for a large-scale public art installation in Melbourne, Australia for the Big West Arts Festival in 2009. The work was vandalized the night of its completion. A new movie, shot by a Tasmanian filmmaker on a set made almost entirely out of yarn, was partially inspired by "knitted graffiti".

CHARITY

Knitting garments for free distribution to others has become common practice among knitting groups. Girls and women knitted socks, sweaters, scarves, mittens, gloves, and hats for soldiers in Crimea, the American Civil War, and the Boer Wars; this practice continued in World War I, World War II and the Korean War, and continues for soldiers in Iraq and Afghanistan. The Australian charity *Wrap with Love* continues to provide blankets knitted by volunteers to people most in need around the world who have been affected by war.

In the historical projects, yarn companies provided patterns approved by the various branches of the armed services; often they were distributed by local chapters of the American Red Cross. Modern projects usually entail the knitting of hats or helmet liners; the liners provided for soldiers must be of 100 per cent worsted weight wool and be crafted using specific colours.

Clothing and afghans are frequently made for children, the elderly, and the economically disadvantaged in various countries. Pine Ridge Indian Reservation accepts donations for the Lakota people in the United States. Prayer shawls, or shawls in which the crafter meditates or says prayers of their faith while knitting with the intent on comforting the recipient, are donated to those experiencing loss or stress. Many knitters today knit and donate "chemo caps," soft caps for cancer patients who lose their hair during chemotherapy. Yarn companies offer free patterns for these caps.

Penguin sweaters were knitted by volunteers for the rehabilitation of penguins contaminated by exposure to oil slicks. The project is now complete.

Chicken sweaters were also knitted to aid battery hens that had lost their feathers. The organization is not currently accepting donations, but maintains a list of volunteers.

Originally started after the 2004 Indonesian tsunami, Knitters Without Borders is a charity challenge issued by knitting personality Stephanie Pearl-McPhee that encourages knitters to donate to Médecins Sans Frontières (Doctors Without Borders). Instead of knitting for charity, knitters are encouraged to donate a week's worth of disposable income, including money that otherwise might have been spent on yarn. Knitted items are occasional offered as prizes to donors. As of September 2011, Knitters Without Borders donors have contributed CAD$1,062,217.

Security blankets can also be made through the Project Linus organization which helps needy children. There are organizations that help reach other countries in need such as afghans for Afghans. This outreach is described as, "afghans for Afghans is a humanitarian and educational people-to-people project that sends hand-knit and crocheted blankets and sweaters, vests, hats, mittens, and socks to the beleaguered people of Afghanistan."

HEALTH BENEFITS

Studies have shown that knitting, along with other forms of needlework, provide several significant health benefits. These studies have found the rhythmic and repetitive action of knitting can "help prevent and manage stress, pain and depression, which in turn strengthens the body's immune system", as well as create a relaxation response in the body which can decrease blood pressure, heart rate, help prevent illness, and have a calming effect. Pain specialists have also found that the brain chemistry is changed when one knits, resulting in an increase in "feel good" hormones (*i.e.* serotonin and dopamine), and a decrease in stress hormones.

Knitting, along with other leisure activities has been linked to reducing the risk of developing Alzheimer's disease and dementia.

Much like physical activity strengthens the body, mental exercise makes the human brain more resilient. A repository of research into the effect on health of knitting can be found at Stitchlinks, an organisation founded in Bath, England.

BOUCLÉ

Bouclé is a kind of novelty yarn. It is a yarn with a length of loops of similar size which can range from tiny circlets to large curls. To make bouclé, at least two strands are combined, with the tension on one strand being much looser than the other as it is being plied, with the loose strand forming the loops and the other strand as the anchor. Bouclé can also refer to the fabric made from this type of yarn, especially fabric that maintains the looped appearance.

MACHINE SPINNING

In machine spinning bouclé can be created in a single step, using a hollow spindle. It is made by differing the feed rate of the effect yarn with respect to the core yarns. The core yarns are wrapped around the effect yarn either loosely or tightly, depending on the difference in feed rates and amount of twist used. The regularity of the loops is controlled by varying the distance between the point where the core yarns come together and the guide. By increasing the distance, the bouclé becomes more irregular.

EYELASH YARN

Eyelash yarn is a type of novelty yarn. It appears as a thread base, with several long strands spaced at even intervals that jut out at an angle from the main strand. The long strands, or hair, can be metallic, opalescent, matte, or a combination of types. The hair can be curly or straight and can sometimes be two different lengths. Prominent types are composed of 100 per cent polyester with a straight and relatively short hair. Because of its thinness, eyelash is normally carried along with another, plainer yarn to add visual interest to the primary yarn. The texture and composition of such yarns have been explored by many companies, and there are innumerable variations of eyelash yarns, such as *pigtail* or *ponytail,* which have a thicker base and what appear to be flags tied onto the base strand at even intervals, or *fur*, in which the base has a more frequently occurring or thicker grouping of hairlike strands which, in the finished fabric, will be hairy and have the general aspect of faux fur.

Some of the drawbacks of these yarns are that they tend to have poor stitch definition because the hairs obscure the appearance of the base and that they add bulk to a garment, so they are mostly used for accessories such as scarves or as garment trim. Some companies have even created eyelash yarn that can be used for decorative purposes for events, such as holidays or other celebrations, by combining metal-look lurex with a polyester core which mimics a tinsel effect. The obvious advantage is that metal-look eyelash yarn is far more durable than typical tinsel and can be put away and used again for annual celebrations.

NOVELTY YARNS

Novelty yarns include a wide variety of yarns made with unusual features, structure or fiber composition such as slubs, inclusions, metallic or synthetic fibers, laddering and varying thickness introduced during production. Some linens, wools to be woven into tweed, and the uneven filaments of some types of silk are allowed to retain their normal irregularities, producing the characteristic uneven surface of the finished fabric. Man-made fibres, which can be modified during production, are especially adaptable for special effects such as crimping and texturizing.

COMPLEX YARNS

Novelty yarns are sometimes referred to as complex yarns. A yarn which makes a fabric is not always smooth and uniform. Complex yarns are the uneven yarns which may be thick and thin or have curls, loops, twists and even differently coloured areas along their length. This look of the yarns are used to add interesting effects in fabrics. In complex ply two or more complex yarns are twisted around each other to form loops, curls and knots to create fancy effects. Many knitting yarns are complex ply yarns which give interesting textures in finished products.

TYPES

- Core: A yarn in which the core has been wrapped by another strand, such as of cotton or nylon around an elastic base as used in commercial socks.

- Eyelash: Eyelash yarn is made from a polyester fiber with a furry texture resembling eyelashes. These novelty yarns are made of a thin central ply surrounded by short "hairs". This yarn differs from "fur" type yarn in that it contains evenly spaced threads at intervals between lengths of bare core thread, whereas fur yarns have an abundance of threads covering the entirety of the core thread. Eyelash yarn comes in a wide range of colours, with the "hairs" sometimes being made of multicoloured or metallic fibers.
- Bouclé: Bouclé, or looped, yarns are created by loosely looping an effect yarn around a base yarn. They can be made of any type of fiber and are usually composed of three plies, or strands, wrapped around each other. The texture is created by spinning one of the three plies more loosely than the other two. Fiber artists who choose to create projects in bouclé yarn must use extra care because if not handled carefully, the loose strand may split and snag on the knitting needles or crochet hook.
- Ribbon: Ribbon is a type of yarn that resembles a ribbon. It can be made from synthetic or natural fibers, such as silk or cotton. Some ribbon yarns are flat, while others are tubular in construction.
- Ladder: Ladder resembles a ladder, with two flat threads representing the two sides of the ladder held together by a strip of material at the center that represents the rungs. The material at the center of ladder yarn can be metallic, beaded, or otherwise adorned. This type of yarn is more often used to create trim or embellishments than to knit or crochet entire garments.
- Nub: A nub or knot is created by tightly twisting an effect fiber around the base fiber. The nub is most easily identified when the effect and base yarns are of different colours.
- Slub: A slub or thick spot in a yarn is created by varying the tightness of the twist of the yarn at various intervals.
- Corkscrew or spiral: The appearance of corkscrew or spiral yarns is achieved by using yarns of two different fibers and often twisting one under a different tension than the other.
- Chenille: The soft, fuzzy surface of chenille yarns, which resemble pipe cleaners in appearance, can be created in several ways. Most commonly, a fabric is first produced and then cut into narrow strips resembling a yarn. Then, when the fabric is cut, the raw edges become very fuzzy and produce the chenille appearance. Other chenilles are created by trimming a loosely attached effect fiber to create the fuzzy appearance. Still other chenilles are created by attaching or gluing fibers to the yarn.

- Metallic yarns: Metallic yarns are often classified as fancy yarns and are created by adding a metallic fiber or yarn to the blend. These are not to be confused with actual wire used in jewelry that is sometimes knit or crocheted.
- Crepe yarns: Crepe yarns may be classified as fancy yarns and are created by tightening the twist given to a yarn, resulting in a kinked or looped strand.

VARIEGATED YARN

Variegated yarn is yarn dyed with more than one colour. It can produce effects that vary depending on the technique of the crafter, the pattern used, and the frequency of colour change. These effects include "flashing" (lightning-bolt effects) and "pooling" (patchy or marbleized effects). Some yarns (known as "self-striping yarns") are designed to produce stripes when used to knit small items such as socks or mittens.Variegated yarn may be used in knitting, crocheting, and other textile arts.

CIRCULAR KNITTING

Circular knitting or knitting in the round is a form of knitting that creates a seamless tube. When knitting circularly, the knitting is cast on and the circle of stitches is joined. Knitting is worked in rounds (the equivalent of rows in flat knitting) in a spiral.

Originally, circular knitting was done using a set of four or five double-pointed needles. Later, circular needles were invented, which can also be used to knit in the round: the circular needle looks like two short knitting needles connected by a cable between them.

Longer circular needles can be used to produce narrow tubes of knitting for socks, mittens, and other items using the Magic Loop technique. Machines also produce circular knitting; double bed machines can be set up to knit on the front bed in one direction then the back bed on the return, creating a knitted tube. Specialized knitting machines for sock-knitting use individual latch-hook needles to make each stitch in a round frame.

Many types of sweaters are traditionally knit in the round. Planned openings (arm holes, necks, cardigan fronts) are temporarily knitted with extra stitches, reinforced if necessary. Then the extra stitches are cut to create the opening, and are stitched with a sewing machine to prevent unraveling. This technique is called steeking.

The Magic Loop technique Invented by Sarah Hauschka and written in Beverly Galeska's booklet "The Magic Loop," published by FiberTrends, the Magic Loop uses a 40-inch-long circular needle to knit projects of any circumference in the round. The key is pulling a loop of extra cable out between the stitches halfway through the round. Using a 47-inch-long circular needle lets the knitter knit two-at-a-time projects, like two socks at once, using the Magic Loop method.

COMBINED KNITTING

Combined knitting or combination knitting is a style that combines elements of Eastern-style knitting with the Western techniques. The name was suggested by Mary Thomas in her 1938 book "Mary Thomas's Knitting Book", where she described the technique as "This is the better way to work in Flat Knitting. The resulting fabric is more even and closer in construction." By wrapping the yarn the opposite way while purling, the knitter changes the orientation of the resulting loops; then the next row's knit stitches can be formed by inserting the needle from the right (as in Eastern knitting), rather than from the left. The needle is always inserted from the right, whether knitting or purling. This technique is suitable for all knitted fabrics from the basic Stockinette stitch, to any other style, such as Fair Isle, circular knitting, or lace knitting.

The basic adaptation necessary is to substitute "ssk" when directed to "k 2 tog", and vice versa, to orient the slant of the decrease correctly. Most American and European knitting patterns are currently not written to accommodate the needs of Combined knitters. The responsibility rests with the individual knitter to have gained sufficient working knowledge of the changes necessary to convert pattern elements before attempting the entire project, in order for the design to be knitted successfully.

Knitting instructors unfamiliar with this technique will encounter difficulties teaching classes with students using this technique. Proper terminology is essential in assisting teachers to provide adequate instruction to these students.

Teachers should familiarize themselves with the works of Annie Modesitt and Anna Zilboorg, among others. Other basic knitting techniques include English knitting (or "right-hand" knitting) and Continental knitting (or "left-hand" knitting).

CONTINENTAL KNITTING

Knitting with the yarn in one's left hand is commonly referred to as Continental knitting, German knitting, European knitting, or left-hand knitting. Unlike English knitting, the yarn is held in the left hand; the motion of bringing the yarn forward with a knitting needle held in the other hand is thus sometimes known as picking. Continental knitting can be done at a greater rate than English knitting, as the stitches are formed closer to the needle points and the yarn has a shorter travel.

Although the general appearance of fabrics produced by Continental and English techniques is the same, there are subtle differences. the stitches are more square and the fabric is closer when knitted Continental style. Mary Thomas estimates that the difference can be the equivalent of two needle sizes.

Other knitting styles include English knitting (aka *right-hand* knitting) and Combined knitting.

HAND MOTIONS

The motion of the right wrist is used to slip the right needle into the loop of the stitch being knitted and 'scoop' or 'hook' the yarn onto the right needle while the left forefinger holds the yarn across the back of the stitch. An alternative method of collecting the yarn involves using the thumb or index finger of the right hand to hold the yarn in place as the new stitch is being pulled out of the loop. This knitting style is often easier to learn for people with crocheting experience, since the way the yarn is held in the left hand is similar to crochet, and the motion of the right hand is similar to the motion seen in crochet, although the knitting needle is held under the palm of the hand. One major difference in the motion of the right wrist is that in crochet the crochet hook may be held more like a pencil; this method of holding the knitting needle like a pencil was briefly made popular around 1900 under the guise of being more ladylike. Nowadays, however, the majority of knitters hold both needles under the palm.

YARN TENSION

The tension in the yarn is controlled by threading the yarn through the fingers of the left hand. Typically, the yarn is looped around the little finger, across the knuckles and around the index finger.

HISTORY

This style originated in continental Europe, specifically recognized in Germany and began spreading in the early nineteenth century to surrounding countries. This is evident in that the Norwegian word for knitting 'binde' gave way to the German word for knitting 'strikke'.

Continental style knitting, being associated with Germany, fell out of favour in English-speaking countries during World War II; its reintroduction in the United States is often credited to Elizabeth Zimmermann.

Since World War II, both continental and English knitting are used in the United States and England. Japanese knitters tend to favor the continental style and Chinese knitters for the most part use the English style. Many other countries generally use continental knitting, such as Portugal, Greece, Turkey, Bolivia, and Peru.

ENGLISH KNITTING

English knitting, also known as right-hand knitting or throwing, is a style of Western knitting where the yarn to be knit into the fabric is carried in the right hand. This style is prevalent throughout the English-speaking world, though it is by no means universal.

Other Western knitting styles include continental knitting (also known as "left-hand knitting") and combined knitting. Despite the names, choice of knitting style has little to do with the handedness of the knitter; plenty of left-handed individuals use the English style, and plenty of right-handed knitters use Continental.

Various non-Western styles also exist, many of which are substantially similar to these, but which twist each stitch, making for a subtly different-looking fabric.

TECHNIQUE

Here, we assume that there are already stitches on the needles, having been cast on previously. The yarn is wrapped around the right hand for tension (in one method, the tail of the yarn is wrapped around the little finger of the right hand for tension, and over the index finger for control—see illustration). The right hand will hold the needle with the most recently knit stitches. (If at the beginning of a row, the right hand will hold the empty needle.) The left hand holds the other needle.

The knit stitch

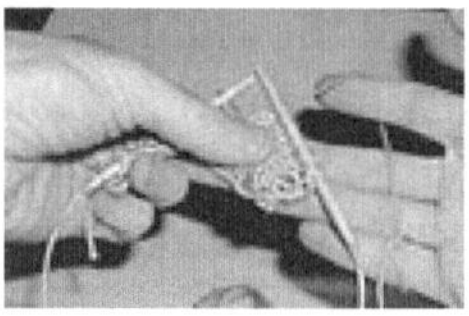

Fig. Holding the yarn

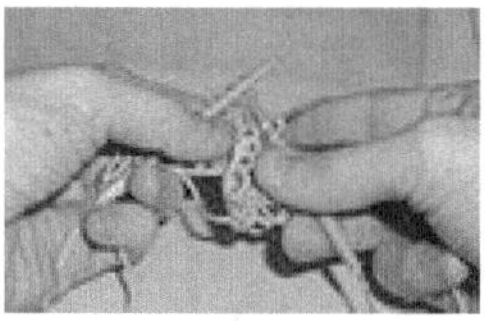

Fig. Placing the needle

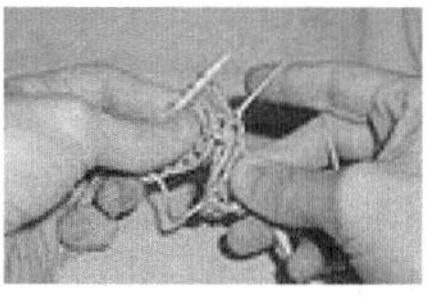

Fig. Inserting the needle

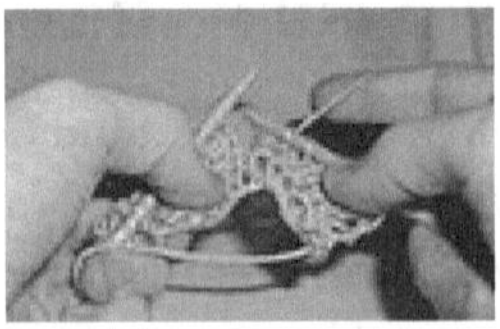

Fig. Opening the loop (exaggerated)

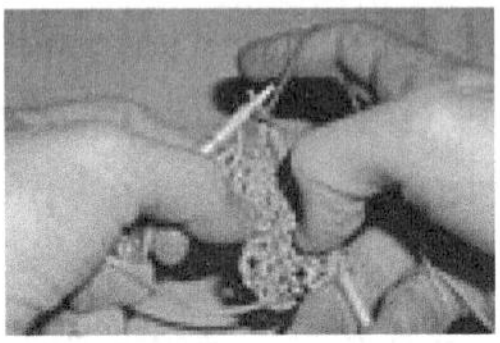

Fig. Wrapping the yarn

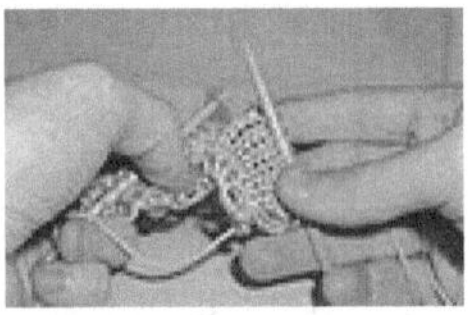

Fig. Pulling the loop through

If the yarn is sitting in front of the right needle (closer to the knitter), it should first be moved between the needles to the back. We will make one knit stitch into the first loop on the left needle. The right needle is inserted into the left side of that loop.

To see what is happening, we can use the two needles to hold that loop wide open: it is through this loop that we will pull the new stitch. The yarn is wrapped *counter-clockwise* around the right needle, and this new loop is pulled with the right needle through the old one. The stitch is now complete. To prepare for the next stitch, we now withdraw the left needle from the just-completed stitch.

The Purl Stitch

If the yarn is sitting behind the right needle (away from the knitter), it should first be moved between the needles to the front. We will make one purl stitch—which looks like the back of a knit stitch—into the first loop on the left needle.

The right needle is inserted into the right side of that loop. Again, to see what is happening, we can use the needles to hold the loop open. Instead of pulling the new loop forward, the right needle is now situated to pull the new loop *backward* through the old loop.

The yarn is still wrapped counter-clockwise around the right needle, and this new loop is then pulled through the old one. The completed stitch is then slid off the left needle.

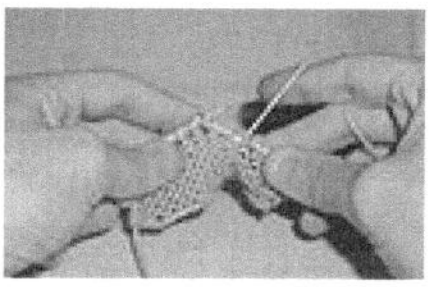

Fig. Inserting the needle

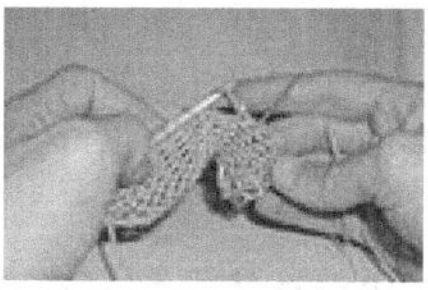

Fig. Opening the loop (exaggerated)

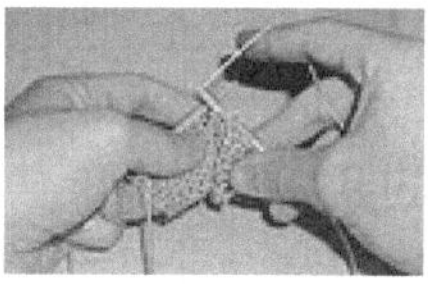

Fig. Wrapping the yarn

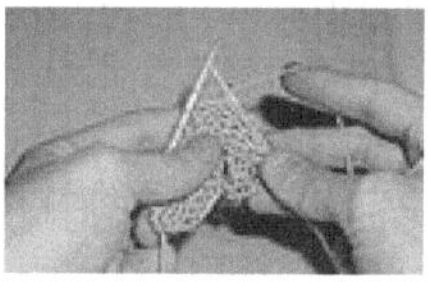

Fig. Drawing the yarn through

FLAT KNITTING

Flat knitting is a method for producing knitted fabrics in which the work is turned periodically, *i.e.*, the fabric is worked with alternating sides facing the knitter. Another method of reaching the same result is to knit alternately from right to left and left to right without turning; this back-and-forth technique requires either innate or learned ambidextrous motor skills.

The two sides (or "faces") of the fabric are usually designated as the right side (the side that faces outwards, towards the viewer and away from the wearer's body) and the wrong side (the side that faces inwards, away from the viewer and towards the wearer's body).

Flat knitting is usually contrasted with circular knitting, in which the fabric is always knitted from the same side. Flat knitting can complicate knitting somewhat compared to circular knitting, since the same stitch is produced by two different movements when knitted from the right and wrong sides. Thus, a knit stitch may be produced by a knit stitch on the right side, or by a purl stitch on the wrong side.

This may cause the gauge of the knitting to vary in alternating rows of stockinette fabrics; however, this effect is usually not noticeable, and may be eliminated with practice (the usual way) or by using needles of two different sizes (an unusual and less effective way).

In flat knitting, the fabric is usually turned after every row. However, in some versions of double knitting with two yarns and double-pointed knitting needles, the fabric may turned after every second row.

In Industrial Knitting applications, the terms "Flat" and "Circular" have very different meanings to those given above. A "Flat" or Vee Bed knitting machine consists of 2 flat needle beds arranged in an upside-down "V" formation.

These needle beds can be up to 2.5 metres wide. A carriage, also known as a Cambox or Head, moves backwards and forwards across these needle beds, working the needles to selectively, knit, tuck or transfer stitches. A flat knitting machine is very flexible, allowing complex stitch designs, shaped knitting and precise width adjustment. It is, however relatively slow when compared to a circular machine. The two largest manufacturers of industrial flat knitting machines are Stoll of Germany, and Shima Seiki of Japan. The industrial hand flat knitting machine is considered to be launched by the Isaac Lamb patents.

WARP KNITTING

Warp knitting is a family of knitting methods in which the yarn zigzags along the length of the fabric, *i.e.*, following adjacent columns ("wales") of knitting, rather than a single row ("course"). For comparison, knitting across the width of the fabric is called weft knitting. Since warp knitting requires that the number of separate strands of yarn ("ends") equals the number of stitches in a row, warp knitting is almost always done by machine, not by hand.

TYPES

Warp knitting comprises several types of knitted fabrics, including tricot, raschel knits, milanese knits and stitch-bonding. All warp-knit fabrics are resistant to runs and relatively easy to sew. Raschel lace—the most common type of lace—is a warp knit fabric but using many more guide-bars (12+) than the usual machines which mostly have three or four bars.

- Tricot is very common in lingerie. The right side of the fabric has fine lengthwise ribs while the reverse has crosswise ribs. The properties of these fabrics include having a soft and 'drapey' texture with some lengthwise stretch and almost no crosswise stretch.
- Milanese is stronger, more stable, smoother and more expensive than tricot and, hence, is used in better lingerie. These knit fabrics are made from two sets of yarn knitted diagonally, which results in the face fabric having a fine vertical rib and the reverse having a diagonal

structure, and results in these fabrics being lightweight, smooth, and run-resistant. Milanese is now virtually obsolete.

- Raschel knits do not stretch significantly and are often bulky; consequently, they are often used as an unlined material for coats, jackets, straight skirts and dresses. These fabrics can be made out of conventional or novelty yarns which allows for interesting textures and designs to be created. The qualities of these fabrics range from "dense and compact to open and lofty [and] can be either stable or stretchy, and single-faced or reversible. The largest outlet for the Raschel Warp Knitting Machine is for lace fabric and trimmings.
- Stitch-Bonding is a special form of warp knitting and is commonly used for the production of composite materials and technical textiles. As a method of production, stitch-bonding is efficient, and is one of the most modern ways to create reinforced textiles and composite materials for industrial use. The advantages of the stitch-bonding process includes its high productivity rate and the scope it offers for functional design of textiles, such as fiber-reinforced plastics. Stitch-bonding involves layers of threads and fabric being joined together with a knitting thread, which creates a layered structure called a multi-ply. This is created through a warp-knitting thread system, which is fixed on the reverse side of the fabric with a sinker loop, and a weft thread layer. A needle with the warp thread passes through the material, which requires the warp and knitting threads to be moving both parallel and perpendicular to the vertical/warp direction of the stitch-bonding machine. Stitch-bonded fabrics are currently being used in such fields as wind energy generation and aviation. Research is currently being conducted into the usage and benefits of stitch-bonded fabrics as a way to reinforce concrete. Fabrics produced with this process offer the potential of using "sensitive fiber materials such as glass and carbon with only little damage, non-crimp fiber orientation and variable distance between threads".
- Needle shift technique is when "Both outer warp layers [are] secured in one procedure by incorporating a shift of the needle bar during the stitching process, creating endless possibilities for the arrangement and patterns in the stitch-bonding process.
- Extended Stitch-Bonding process (or the extended warp-knitting process): the compound needle that pierces the piles is shifted laterally according to the yarn guides. This then makes it possible for the layers of the stitch-bonded fabric to be arranged freely and be made symmetrical in one working step. This process is advantageous to the characteristics of the composite as the "residual stresses

resulting from asymmetric alignment of the layers are avoided, [while] the tensile strength and the impact strength of the composite are improved".

ADVANTAGES

Producing textiles through the warp knitting process has the following advantages

- Higher productivity rates than weaving
- Variety of fabric constructions
- Large working widths
- low stress rate on the yarn that allows for use of fibers such as glass, aramide and carbon
- the creation of three-dimensional structures that can be knitted on double needle bar raschels

APPLICATIONS

Warp knitted fabrics have several industrial uses, including producing mosquito netting, tulle fabrics, sports wear, shoe fabric, fabrics for printing and advertising, coating substrates and laminating backgrounds. Research is also being conducted into the use of warp knitted fabrics for industrial applications (for example, to reinforce concrete), and for the production of biotextiles.

WARP KNITTING AND BIOTEXTILES

The warp knitting process is also being used to create biotextiles. For example, a warp knitted polyester cardiac support device has been created to attempt to limit the growth of diseased hearts by being installed tightly around the diseased heart. Current research on animals "have confirmed that...the implantation of the device reverses the disease state, which makes this an alternative innovative therapy for patients who have side effects from traditional drug regimes".

COATS PLC

Coats plc is an industrial thread and consumer textile crafts business.

In the 1750s James and Patrick Clark began work in the loom equipment and silk thread business in Paisley, Scotland In 1806 Patrick Clark invented a way of twisting cotton threads together to substitute for silk threads which were unavailable due to France's blockade of Great Britain and opened the first plant for manufacturing the cotton thread in 1812.

In 1802 James Coats set up a weaving business, also in Paisley. In 1826 he opened a cotton mill at Ferguslie to produce his own thread and, when he retired in 1830, his sons, James and Peter, took up the business under the name of J.

and P. Coats. The firm expanded internationally, particularly to the USA. In 1890 Coats listed on the London Stock Exchange, with a capital base of £5.7M.

In 1952 J. and P. Coats and the Clark Thread Co. merged. In 1961 a merger with Patons and Baldwins created Coats Patons. In 1986 a merger with Vantona Viyella created Coats Viyella. In 2003 the Guinness Peat Group took over Coats Ltd and it is now registered as "Coats plc".

AMERICAN BRANCH

Coats began selling its thread in the United States by the 1830s. It set up a network of agencies to sell its own Coats brand and, in 1869, began local manufacture of thread in Rhode Island. In 1864 the Clark family began manufacturing in Newark, New Jersey as the Clark Thread Co. Thanks to its well-timed direct investments and the development of a flexible managerial structure, Coats came to dominate the American cotton thread market.

SPANISH BRANCH

In 1903 J andP Coats joined a local company (*Sucesores de Fabra y Portrabella*) creating the *Compañía anónima de Hilaturas Fabra y Coats*, that became leader of the Spanish thread market as well as one of the main textile industries in Spain. Fabra y Coats had several factories in Barcelona and its surrounding area. The most important was the existing in the Sant Andreu district which activity ceased in 2005, though its industrial buildings have been preserved and currently house social and community services.

OPERATIONS

Coats plc does business under different names in different countries but most variations include "Coats". For example, in the U.S., it does business as "Coats and Clark, Inc." (consumer) and "Coats American, Inc." (industrial). In the UK and Ireland, it does business as "Coats (UK) Ltd", in Canada as "Coats Canada Inc" (and has an industrial subsidiary "Coats Bell"), and as "Coats Australian Pty Ltd" in Australia. Coats in Egypt does business as "Coats Egypt". It was also operating in India as "Madura Coats" famous for its "9to5 Man" brand of textiles.

CONTROVERSY

In June 2012, Coats plc lost court challenges against European Union antitrust fines for fixing the price of fastners such as zips, rivets and snaps. The European General Court dismissed Coats' appeal on all counts and upheld the 110m Euro fine imposed by the European Commission in its original decision of 19 September 2007. In February 2012, it is alleged union leaders were held captive by factory managers in Bangladesh whilst trying to resolve an industrial dispute. Union leaders had their mobile phones confiscated and were forcibly detained in Coats' head office building overnight, only releasing them when trade union colleagues alerted the local media.

PRODUCTS

Coats makes and sells thread, zips, yarn, speciality textiles and related goods. Recent innovations have included Coats Protect, the world's first antimicrobial sewing thread Coats Colour Express, a revolutionary web-based service that provides the apparel and footwear industry with the world's fastest and most colour accurate sampling facility for industrial sewing thread and Coats Insectiban, the world's first anti-bed bug sewing thread and zips

- Speciality: Coats sells items catering to the needs of industries such as footwear, mattresses, quilting, automotive, outdoor and camping goods and furniture. Their operational units are accredited with TS16949, ISO9001, Ford Q1 Mark, SATRA Quality Mark and others.
- Zips: Coats produces zips for industrial applications in the apparel and speciality sectors, as well as for craft and home sewing usage.
- Crafts: Coats make products suitable for a variety of needlecrafts such as knitting, crochet, embroidery and quilting.

COATS LAND

It's over a hundred years since the discovery of Coats Land was made in the Antarctic, funded primarily by Major Andrew Coats and his younger brother James Jnr. from the famous Paisley thread manufacturing family. They provided over £30,000 of the £36,405 raised to finance the venture and were staunch supporters of the expedition, also providing a great number of Fair Isle jerseys, caps, gloves, scarves and stockings which added a splash of colour to the Antarctic scene long before they became fashion items.

In December 2012 it was announced, during a historic visit to Downing Street by the Queen, that part of Antarctica had been named Queen Elizabeth II Land as a Diamond Jubilee gift from the Foreign Office.

To commemorate the naming of Queen Elizabeth II Land and mark Her Majesty's Jubilee year, Coats has created a two by three metre knitted British flag designed by Rowan Head Designer Marie Wallin using Rowan Purelife British Sheep Breeds yarns. Rowan is a strong supporter of the Campaign for Wool and Love Wool UK. The Campaign for Wool is a global community of sheep farmers, retailers, designers and manufacturers, united by patron His Royal Highness The Prince of Wales. The Campaign for Wool aims to educate as many people as possible about the benefits and versatility of using wool in fashion, furnishings and everyday life; supporting many small businesses and local farmers whose livelihoods depend on the wool industry.

LION BRAND

Founded in 1878, Lion Brand Yarn Company is the oldest manufacturer of knitting and craft yarn in the United States. The company has a storefront in Manhattan, publishes several knitting and crochet newsletters, and is involved in charitable works.

Lion Brand Yarns have been on the market since 1878. Current President/ CEO David Blumenthal's great-grandfather Reuben founded this 5th generation, family-owned company.

Company was founded in 1878 by a group of notions and dry goods salesmen that included current CEO David Blumenthal's great-grandfather, Reuben. During the 1930s under the leadership of Reuben, Lion Brand imported yarns from Europe, supplementing production rather than manufacturing yarns in the U.S. This was considered to be an unusual business strategy at the time. The company currently produces yarn in the U.S. and all over the world.

Joseph Blumenthal, Rueben's son-in-law, had three sons who ran the company from the 1940s through the early 1990s. Isidor Blumenthal (b1909, d 2003) served as president of the company from 1958 to his death in 2003. George Blumenthal was in charge of sales and started selling to major discount chains, including Target, in 1962. Bernard Blumenthal ran logistics and distribution.

The fourth generation consisted of Isidor's son, David, currently President and CEO, who began working for Lion Brand on October 1, 1969, and his first cousins, Dean (COO/EVP) who opened the Wal Mart account in 1983, Jack, Senior VP, who travels as an ambassador, speaking to groups around the country about the business, and oversees sales. and Alan, Senior VP who is involved in logistics and distribution. Several fifth generation Blumenthals work at Lion Brand as well, Adam - Director of Corporate Strategy, Shira - Studio Associate, Evan - Director Global Business Development and David, who joined the family business in 2014.

The company is privately held and has estimated sales of $100 million per year.

- Vanna White: In 1992 Johnny Carson interviewed Vanna White on *The Tonight Show* where she disclosed her favorite hobby was crocheting. David Blumenthal's cousin, Jack heard the interview on TV, which resulted in David sending White a package of Lion Brand Yarn. White then began working as a spokesperson for Lion Brand which continues today. Lion Brand carries a variety of yarn called "Vanna's Choice". Additional yarns to White's line include "Vanna's Glamour", "Vanna's Sequins" and "Vanna's Choice Baby." A portion of the proceeds from sales of these yarns are donated to St. Jude Children's Research Hospital. On the June 10, 2013 broadcast of

Wheel of Fortune (U.S. game show) the programme showcased White presenting a check along with David and Dean Blumenthal on behalf of Lion Brand in the amount of $1,000,000.00 to St. Jude's. Game show host, Pat Sajak noted that St. Jude's life saving mission is a "great cause" and a "great charity."

- Ernst and Young Entrepreneur Of The Year Award : In June 2013, it was announced that Lion Brand Yarn's President/CEO, David Blumenthal and EVP/COO, Dean Blumenthal were awarded Ernst and Young Entrepreneur Of The Year Award for their entrepreneurial excellence in the 'Family Business' category.
- Publishing : Lion Brand has published several books featuring patterns using their yarn. *Lion Brand Yarn: Vintage Styles for Today* has dozens of patterns that are Vintage in flair, which have been adapted from original patterns from the Lion Brand archives. The patterns have also been updated to feature many types of yarn produced by Lion Brand today. Lion Brand has partnered with several publishers to provide patterns using their products. Some of these include: Knit.1 Magazine published by Vogue Knitting, Clarkson Potter which is part of Crown Publishing Group, Leisure Arts and Random House.
- Web site : Lion Brand's official web site is the one of the most visited yarn information web sites and has over 5,000 free knitting, crochet and craft patterns. Its weekly e-newsletter, The Weekly Stitch, has over 1 million subscribers. Additionally, Lion Brand has a strong digital presence through its daily activity on social media platforms such as Twitter, Pinterest and Facebook. In 2006, Lion Brand won a Webby Award in the Corporate Communications category and its podcast, titled "YarnCraft," has won several awards including PR News' best podcast/videocast award in 2008.
- Free Patterns: Lion Brand offers over 5,000 free knitting, crochet and craft patterns.
- Yarn: Lion Brand Yarn Company offers a variety of yarns, from 100 per cent natural fibers including cotton, cashmere, alpaca and wool to blends and 100 per cent acrylics in a variety of weights. Some of their most popular yarns are *Homespun, Fun Fur, Vanna's Choice,* and *Wool-Ease*. The company came out with organic cotton yarn in 2007.
- Lion Brand Yarn Studio: The Lion Brand Yarn Studio is a retail store located in Manhattan, which opened in 2008. The Studio sells yarn and provides knitting and crochet classes. The Studio also offers custom knitting and crochet services. Since the Studio opened, they're taught over 12,000 students.

- Lion Brand Yarn Outlet: The Lion Brand Yarn Outlet is a retail outlet store located in Carlstadt, New Jersey, which opened in 2011. The Outlet sells discontinued yarns and provides a lounge for customers to knit and crochet.

LOPI

Lopi is knitting wool made from the fleece of Icelandic sheep. The fleece is made up of two layers, each with a different kind of wool. The wet-resistant outer coat contains long, coarse fibres, while the insulating layer beneath consists of soft, short fibres. These are processed together to create lopi roving and yarn. The machine-carded roving is produced in disc-shaped rolls. This is the original unspun lopi first used for knitting around 1900.

More recently, lightly spun lopi yarn in different thicknesses has been available. Most wool produced in Iceland is processed by Ístex, the Icelandic Textile Company. They manufacture 3 types of spun lopi yarn and also unspun lopi. It comes in a variety of natural fleece shades, and in a range of dyed colours. Lopi from local producers is also available, for instance in Þingborg near Selfoss. Characteristic Icelandic lopapeysa sweaters are generally made from the thicker lopi yarns.

EISAKU NORO COMPANY

Eisaku Noro Company, Ltd. is a yarn manufacturer located in the Aichi Prefecture of Japan. It produces yarns for handcrafting under the Noro brand name, as well as machine yarns for textile production using the Eisaku Noro label. The company was founded over forty years ago by Eisaku Noro. The handcrafting yarns in particular are well known for their vivid colours and combinations of diverse fiber types.

They differ from a number of other manufactured yarns in the industry by having lengthier spans of colour in the runs, causing distinctive striping patterns, as well as being partially spun by hand versus being produced completely by machines.

PRODUCTION

All animal fibers that are used in the yarns come from certified organic farms, from international sources such as Australia, the Falkland Islands and South Africa. Noro is personally involved with inspecting all aspects of production, from visiting the animal farms to checking the machinery used and keeping restrictions on the dye processes to maintain products that are as eco-friendly as possible.

He says "Friction, rubbing and heat during processing weaken the fibers in direct proportion to the length of time they are processed. By dramatically shortening this process, we are preventing damage to the enzymes in the fibers and simultaneously profiting the environment."

Differences in processing from other yarn manufacturers include the use of a single roller on the carding machine instead of the usual three, hand-feeding tufts of fiber into the carding machine, and the arrangement of specified colour runs by hand. The single roller as well as the shorter bed on the spinning machine result in less twist to the single ply than other yarns, giving a variance in thickness along its length.

2011 EARTHQUAKE

Following the devastating 2011 Tôhoku earthquake and tsunami, Eisaku Noro addressed a letter on March 18, 2011 to their US distributor, Knitting Fever, and to his customers, who desired to donate funds for relief efforts for Japan. The letter states that he and his company were fortunate not to have suffered any damages, relates the plight of the people who were affected, and expresses his gratitude for those concerned for them.

Another letter, from Noro's son and operations manager Takuo Noro, on the European handcraft yarn distributor's web site, gives details of the distances of their main site and dye house from the damaged Fukushima nuclear power plant, as well as the ports their company uses to import raw goods and export products. The ports, in Nagoya, Osaka, and Kobe, are located to the south or west of the company.

He also states that the production facilities are all indoors and the radiation levels in these areas have not changed from normal levels. He concludes the letter saying there have been no changes in their surroundings since the earthquake and that the Japanese government was considering conducting radiation inspections on all of the nation's outgoing products.

ARAN JUMPER

The Aran jumper (Irish Gaelic: *Geansaí Árann*) is a style of jumper that takes its name from the Aran Islands off the west coast of Ireland. A traditional Aran Sweater usually is off-white in colour similar to a sheep's wool and is made from 100 per cent wool.

One of the most recognisable features of a tradition Aran is its cable patterns on the chest. Being made of 100 per cent wool it can retain over 30 per cent of its weight in moisture. Nowadays Lanolin is removed from Aran jumper but this waxy substance secreted by sheep in this wool gave the jumper its all-weather versatility.

Originally the jumpers were knitted using unscoured wool that retained its natural oils (lanolin) which made the garments water-resistant and meant they remained wearable even when wet. It was primarily the wives of the islander who knitted the jumpers before local knitters began selling their produce through initiatives such as Congested Districts Board for Ireland in 1892.

The jumpers are distinguished by their use of complex textured stitch patterns, several of which are combined in the creation of a single garment. The jumper (or indeed other options such as "pullover" and "jersey") is largely determined by the regional version of English being spoken. In the case of Ireland and Britain and Australia, "jumper" is the standard word with "sweater" mainly found in tourist shops and America. The word used in Irish is *geansaí*, a Gaelicisation of *guernsey* which has been re-Anglicised to *gansey* in Hiberno-English.

Some stitch patterns have a traditional interpretation, often of religious significance. The honeycomb is a symbol of the hard-working bee. The cable, an integral part of the Aran islander's daily life, is said to be a wish for safety and good luck when fishing. The diamond is a wish of success, wealth and treasure. The basket stitch represents the basket, a hope for a plentiful catch.

CHARACTERISTICS

Traditionally, an Aran jumper is made from undyed cream-coloured *báinín* (pronounced "bawneen"), a yarn made from sheep's wool, sometimes "black-sheep" wool. They were originally made with unwashed wool that still contained natural sheep lanolin, making the garment extent water-repellent. Up to the 1970s, the island women spun their own yarn on spinning wheels.

The jumper usually features 4–6 texture patterns each of which is about 5–10 cm (2–4 in) in width, that move down the jumper in columns from top to bottom. Usually, the patterns are symmetrical to a centre axis extending down the centre of the front and back panel. The patterns also usually extend down the sleeves as well. The same textured knitting is also used to make socks, hats, vests, and skirts.

HISTORY

There is debate about when island residents first started making the jumpers. Undoubtedly, residents of the islands produced a local version of a gansey jumper similar to other areas of the British Isles for several centuries. Traditional ganseys from neighbouring regions have much of the same cabling and pattern-work seen in Aran jumpers; however these ganseys use different construction methods and are knit from a finer wool.

Some have suggested that the jumper is an ancient design that has been used on the islands for hundreds of years. Proponents of this theory often point to a picture in the Book of Kells that appears to depict an ancient "Aran jumper". Also, many megaliths around Europe depict similar patterns to those used in the knitting, which are carved into the stone, and date back several thousand years. However, it is more likely that the knitting stitches were modelled on these than that they evolved contemporaneously.

Most historians agree that far from being an ancient craft, Aran knitting was invented as recently as the early 1900s by a small group of enterprising island women, with the intention of creating garments not just for their families to wear but which could be sold as a source of income. These women adapted the traditional gansey jumper by knitting with thicker wool and modifying the construction to decrease labour and increase productivity.

The first commercially available Aran knitting patterns were published in the 1940s by Patons of England. *Vogue* magazine carried articles on the garment in the 1950s, and jumper exports from the west of Ireland to the United States began in the early 1950s. Standun in Spiddal, Co.Galway was the first to export the Aran sweater to the USA. This provided employment for women throughout Ireland.

The development of the export trade during the 1950s and 1960s took place after P.A. Ó Síocháin organised an instructor, with the help of a grant from the Congested Districts Board for Ireland, to go to the islands and teach the knitters how to make garments to standard international sizings. He commissioned the Irish artist Seán Keating, who had spent much time on the islands, to design and illustrate marketing brochures.

Knitting became an important part of the islands' economy. Adding to the popularity of the Aran jumpers were The Clancy Brothers and Tommy Makem, an Irish folk music group which started recording in New York City in the late 1950s and who adopted the Arans as their trademark on-stage garments. In the early 1960s they appeared on *The Ed Sullivan Show* and even in a special televised performance for US President John F. Kennedy.

The national exposure and the rising popularity of The Clancy Brothers and Tommy Makem fuelled the demand for Aran jumpers even more. During the 1960s, even with all available knitters recruited from the three islands and from other parts of Ireland, Ó Síocháin had difficulty in fulfilling orders from around the world.

Myths

Part of the appeal and popularity of Aran Jumper comes from the array of myths surrounding the garment.

1. The moss stitch is said to signify an abundance of growth
2. The blackberry stitch represents nature.
3. The honeycomb is a said to be a lucky stitch, signifying plenty
4. Lattice or basket stitches to represent the old wicker basket patterns.
5. The Ladder of Life and Tree of Life represent the stages of life.

It is sometimes said that each Islander (or his family) had a jumpers with a unique design, so that if he drowned and was found, maybe weeks later, on the beach, his body could be identified. This misconception may have originated with J.M. Synge's 1904 play *Riders to the Sea,* in which the body of a dead Islander is identified by the hand-knitted stitches on one of his garments.

However, even in the play, there is no reference to any decorative or Aran-type pattern. The garment referred to is a plain stocking and it is identified by the number of stitches, the quote being "it's the second one of the third pair I knitted, and I put up three score stitches, and I dropped four of them". There is no record of any such event ever having taken place, nor is there any evidence to support there being a systematic tradition of family patterns.

There is, however, a long-standing tradition of jumper patterns having a regional or local identification. It is said that the county, or parish, or township of a sailor or an Islander could be identified by his jumper pattern. Additionally, the wearer's initials were traditionally knit into the bottom of the garment, which would have been a far better indication of identity than the stitching pattern.

ARAN PRODUCTION TODAY

While in the past, the majority of jumpers and other Aran garments were knitted by hand, today the majority of items for sale in Ireland and elsewhere are either machine knit or produced on a hand loom. There are very few people still knitting jumpers by hand on a commercial basis but hand knit jumpers are still available at local craft initiatives.

Machine-knitted jumpers tend to use finer wool and have less complex patterns, since many of the traditional stitches cannot be reproduced this way. They are generally flatter, lighter and less substantial, without the pronounced texture of a hand knitted or hand loomed jumper. Machine knitted jumpers will loosen with wear, and so it is advisable to buy one size smaller than the size of the sweater usually worn. They are the least expensive option. Hand-looming allows more complicated stitches to be used, will have fewer stitches to the inch and be thicker.

The best quality hand-loomed jumpers are almost indistinguishable from hand knit. Hand-knit jumpers tend to be more tightly knit, to have more complex stitch patterns and to be longer-lasting and they attract a significant price premium. By holding them up to light, the difference between the machine knit and hand knit is evident.

ARGYLE (PATTERN)

The argyle (occasionally argyll) pattern is made of diamonds or lozenges. The word is sometimes used to refer to an individual diamond in the design but more commonly refers to the overall pattern. Most argyle layouts contain layers of overlapping motifs, adding a sense of three-dimensionality, movement, and texture. Typically, there is an overlay of intercrossing diagonal lines on solid diamonds.

The argyle pattern is derived from the tartan of Clan Campbell, of Argyll in western Scotland, used for kilts and plaids, and from the patterned socks worn by Scottish Highlanders since at least the 17th century. (See illustrations in History of the kilt) These were generally known as "tartan hose".

Argyle knitwear became fashionable in Great Britain and then the USA after the first world war.

Pringle of Scotland popularised the design, helped by its identification with the Duke of Windsor. Pringle's web site says that "the iconic Pringle argyle design was developed" in the 1920s. The duke, like others, used this pattern for golf clothing: both for jerseys and for the long socks needed for the plus-fours trouser fashion of the day.

Payne Stewart (1957–1999), who won the U.S. Open and a PGA championship, was known and loved by fans for his bright and "flashy" dress; he wore tams, knickerbockers, and argyle socks.

As a knitting pattern, argyle is generally accomplished using the *intarsia* technique. Argyle patterns are occasionally woven.

Some sports teams use bright, contemporary interpretations of the argyle pattern: for example, the Garmin-Slipstream professional cycling team, nicknamed the "Argyle Armada", and the Norwegian men's curling team at the 2010 Winter Olympics.

On April 27, 2013 the professional soccer team Sporting Kansas City of Major League Soccer (MLS) in the United States announced their third kit of the 2013 season featuring an argyle pattern.

FAIR ISLE (TECHNIQUE)

Fair Isle is a traditional knitting technique used to create patterns with multiple colours. It is named after Fair Isle, a tiny island in the north of Scotland, that forms part of the Shetland islands.

Fair Isle knitting gained a considerable popularity when the Prince of Wales (later to become Edward VIII) wore Fair Isle tank tops in public in 1921. Traditional Fair Isle patterns have a limited palette of five or so colours, use only two colours per row, are worked in the round, and limit the length of a run of any particular colour.

Some people use the term "Fair Isle" to refer to any colourwork knitting where stitches are knit alternately in various colours, with the unused colours stranded across the back of the work. Others use the term "stranded colourwork" for the generic technique, and reserve the term "Fair Isle" for the characteristic patterns of the Shetland Islands. Other techniques for knitting in colour include intarsia, slip-stitch colour (also known as mosaic knitting).

TECHNIQUE

Basic two-colour Fair Isle requires no new techniques beyond the basic knit stitch. (The purl stitch is not used if circular knitting needles or 3 double-pointed needles are used.) At each knit stitch, there are two available "active" colours of yarn; one is drawn through to make the knit stitch, and the other is simply held behind the piece, carried as a loose strand of yarn behind the just-made stitch.

To avoid "loose" strands larger than 3-5 stitches, the yarn not in use can be "caught" by the yarn in use without this being seen on the front of the work. Knitters who are comfortable with both English style and Continental style knitting can carry one colour with their right hand and one with their left, although it is also possible to simply use two different fingers for the two colours of yarn and knit both using the same style.

The simplest Fair Isle pattern uses circular or double pointed needles, cast on any number of stitches. Knitting then continues round and round, with the colours alternated every stitch. If the pattern is started with an even number of stitches, a vertically striped tube of fabric will be formed, while an odd number will create a diagonal grid that appears to mix the two colours.

Traditional Fair Isle patterns normally had no more than two or three consecutive stitches of any given colour, because they were *stranded*, and too many consecutive stitches of one colour means a very long strand of the other, quite easy to catch with a finger or button. A more modern variation is *woven* Fair Isle, where the unused strand is held in slightly different positions relative to the needles and thereby woven into the fabric, still invisible from the front, but trapped closely against the back of the piece. This permits a nearly limitless variety of patterns with considerably larger blocks of colour.

Traditionally Fair Isle jumper construction involves knitting the body of the jumper completely in the round. *Steeks* are worked across the armhole openings allowing the body to be completed in the round without interruption. Once the main body of the jumper is complete, the armhole steeks are cut open (sometimes these are secured before cutting). Stitches are then picked up around the armhole opening and the sleeve is knitted down towards the cuff in the round. ("Steek" from the Scottish word meaning 'stitch', 'to close shut'. A knitted steek is formed of several stitches.

The technique was first developed in the Shetland Isles.) Beginning in the 1990s, the term "Fair Isle" has been applied very generally and loosely to any stranded colour knitting regardless of its relation to the knitting of Fair Isle or any of the other Shetland Islands.

INTARSIA (KNITTING)

Intarsia is a knitting technique used to create patterns with multiple colours. As with the woodworking technique of the same name, fields of different colours and materials appear to be inlaid in one another, but are in fact all separate pieces, fit together like a jigsaw puzzle.

Unlike other multicolour techniques (including Fair Isle, slip-stitch colour, and double knitting), there is only one "active" colour on any given stitch, and yarn is not carried across the back of the work; when a colour changes on a given row, the old yarn is left hanging. This means that any intarsia piece is topologically several disjoint columns of colour; a simple blue circle on a white background involves one column of blue and two of white—one for the left and one for the right. Intarsia is most often worked flat, rather than in the round. However, it is possible to knit intarsia in circular knitting using particular techniques.

Common examples of intarsia include sweaters with large, solid-colour features like fruits, flowers, or geometric shapes. Argyle socks and sweaters are normally done in intarsia, although the thin diagonal lines are often overlaid in a later step, using Swiss darning or sometimes just a simple backstitch.

TECHNIQUE

Knitting in intarsia theoretically requires no additional skills beyond being generally comfortable with the basic knit and purl stitches. Materials required include multiple colours of yarn, standard needles, and *bobbins*. Bobbins serve to contain the inactive yarn and help keep it from getting tangled. Unlike the narrow, wooden ones used to make bobbin lace, modern intarsia bobbins resemble translucent plastic yo-yos that can snap tight to prevent the yarn from unwinding.

After winding a few yards of each colour onto its own bobbin (and possibly several bobbins' worth of some colours), the knitter simply begins knitting their pattern.

When they arrive at a point where the colour changes, the knitter brings the new colour up underneath the old one (to prevent holes) and starts knitting with it. If flat knitting, at the end of the row, the piece is turned round just as with regular knitting, and the knitter returns the way they came.

The simplest intarsia pattern is for straight vertical stripes. After the first row, the pattern is continued by always working each stitch in the same colour as the previous row, changing colours at exactly the same point in each row. To make more elaborate patterns, one can let this colour boundary drift from row to row, changing colours a few stitches earlier or later each time.

Intarsia patterns are almost always given as charts (which, because of the mechanics of knitting, are read beginning at the lower right and continuing upward). The charts generally look like highly pixellated cartoon drawings, in this sense resembling dot-matrix computer graphics or needlepoint patterns (though usually without the colour nuance of the latter).

KNITTING INTARSIA IN THE ROUND

There are several techniques to execute intarsia when circular knitting. Executing intarsia "in-the-round" eliminates the need for seams and, in one case, turning the piece and working from the wrong side. Priscilla Gibson-Roberts discusses four techniques for knitting intarsia in the round in her book, *Ethnic Socks and Stockings*.

3

Stitches

BASIC KNITTED FABRICS

Basic knitted fabrics are so fundamental that some types have been adopted as part of the language of knitting, similar to techniques such as yarn over or decrease. Examples include stockinette stitch, reverse stockinette stitch, garter stitch, seed stitch, faggoting, and tricot.

STOCKINETTE/STOCKING STITCH AND REVERSE STOCKINETTE STITCH

Stockinette stitch (in UK, Australia, New Zealand etc, Stocking Stitch) is the most basic knitted fabric; every stitch is a knit stitch. In the round, stockinette stitch is produced by knitting every stitch; by contrast, in the flat, stockinette stitch is produced by knitting and purling alternate rows. Stockinette-stitch fabric is very smooth and each column ("wale") resembles a stacked set of "V"'s. It has a strong tendency to curl horizontally and vertically because of the asymmetry of its faces.

Reverse stockinette stitch is produced in the same way as stockinette, except that the purl stitches are done on the right side and the knit stitches on the wrong side. In the round, reverse stockinette stitch is produced by purling every stitch.

GARTER STITCH

Garter stitch, also known as the Knit stitch, is the most basic form of welting. In the round, garter stitch is produced by knitting and purling alternate rows. By contrast, in the flat, garter stitch is produced by knitting every stitch (or purling every stitch, though this is much less common).

In garter-stitch fabrics, the "purl" rows stand out from the "knit" rows, which provides the basis for shadow knitting. Garter-stitch fabric has significant lengthwise elasticity and little tendency to curl, due to the symmetry of its faces.

SEED/MOSS STITCH

Seed stitch (called Moss stitch in UK, Australia, New Zealand etc.) is the most basic form of a basketweave pattern; knit and purl stitches alternate in every column ("wale") *and* every row ("course"). In other words, every knit stitch is flanked on all four sides (left and right, top and bottom) by purl stitches, and vice versa.

Seed/moss-stitch fabrics lie flat; the symmetry of their two faces prevents them from curling to one side or the other. Hence, it makes an excellent choice for edging, *e.g.*, the central edges of a cardigan. However, seed stitch is "nubbly", not nearly as smooth as stockinette/stocking stitch.

FAGGOTING

Faggoting is a variation of lace knitting, in which every stitch is a yarn over or a decrease. There are several types of faggoting, but all are an extremely open lace similar to netting. Like most lace fabrics, faggoting has little structural strength and deforms easily; so it has little tendency to curl despite being asymmetrical. Faggoting is stretchy and open, and most faggoting stitches look the same on both sides, making them ideal for garments like lacy scarves or stockings.

TRICOT KNITTING

Tricot is a special case of warp knitting, in which the yarn zigzags vertically, following a single *column* ("wale") of knitting, rather than a single *row* ("course"), as is customary. Tricot and its relatives are very resistant to runs, and are commonly used in lingerie.

DECREASE (KNITTING)

A decrease in knitting is a reduction in the number of stitches, usually accomplished by suspending the stitch to be decreased from another existing stitch or by knitting it together with another stitch.

METHODS OF SINGLE DECREASING (KNITTING)

When more than one stitch is suspended from a stitch, they can hang in different orders. For example, the first stitch could be on top of the second stitch or the reverse, leaning to the left or the right. The order of stitches is important, both for appearance and for the way it pulls the fabric.

- K2tog ("knit two together") — Work to the two stitches to be decreased, insert the right-hand needle into the first two stitches as if to knit, wrap yarn around needle in normal manner, slip the two stitches off together and drop them. This creates a right-leaning decrease.

- SSK ("slip, slip, knit") — Work to the two stitches to be decreased, slip two stitches one at a time to the right-hand needle, as if to knit; insert the left-hand needle into the two stitches from front to back, knit the two stitches together and drop them. This creates a left-leaning decrease.
- S1, K1, PSSO ("slip one, knit one, pass slipped stitch over") — This results in a similar look to the SSK but can appear less tidy. Work to the two stitches to be decreased, slip next stitch to the right-hand needle as if to knit, knit next stitch, pass slipped stitch over knit stitch. Also creates a left-leaning decrease.
- K2tog tbl ("knit two together through the back loops") — This looks similar to the SSK or S1, K, PSSO, but is faster to work. Work to the two stitches to be decreased, then insert the right hand needle into the backs of the next two stitches (*i.e.* behind the left-hand needle, such that the two needles are anti-parallel in the stitches). Wrap the yarn normally and slip the two stitches off the left needle. This makes a left-leaning decrease.

METHODS OF DOUBLE DECREASING (KNITTING)

Sometimes a double decrease is made, in which three stitches are suspended from a single stitch. This allows for six possible stitch orders: 123, 132, 213, 231, 312 and 321. Here, the first number is the topmost stitch, and the last number is the bottommost stitch. Thus, 213 means that the second stitch is uppermost, followed by the first, then third stitches. The uppermost stitch is most important; there is not much visual difference between 213 and 231.

The simplest double decreases are k3tog and p3tog, which both slant to the right. An attractively symmetric double decrease is 213, which can be done as follows: slip stitches 1 and 2 knitwise simultaneously, knit stitch 3, then pass the slipped stitches over the just-knitted stitch.

RELATION TO BINDING OFF

Binding off is effectively a series of adjacent decreases. The simplest binding off method is to pass each knitted loop over the loop next to it. The final loop is secured by passing the knitting yarn through it, so it is best to start at the point furthest from the knitting yarn. This makes a tight edge, in contrast to other binding off methods that have a tendency to flare out. This method also does not require that the knitting yarn be nearby, so it can be done at any time or position, *e.g.*, to form button holes.

The next simplest binding off is successive k2tog or p2tog stitches, or their counterparts k2tog tbl and p2tog tbl. In all these cases, the knitted stitch is returned to the left needle, to be combined with the following stitch.

ROLE IN FABRIC SHAPING

Decreases are useful in shaping the edges of knitted pieces, and also in creating surface curvature in pieces, *e.g.*, by creating darts.

DIP STITCH

In knitting, a dip stitch is made by knitting into a stitch (or even the space between stitches) of an earlier row. The most common type of dip stitch is to knit into the row below, which may be used for visual effect or to increase the number of stitches (a lifted increase). Knitting into the row below and dropping the stitch above is a quick alternative to brioche knitting, which is normally done by knitting together a yarn over and a slip stitch. Longer dip stitches can be made for visual effects; they draw a narrow V-shaped pair of yarns from any earlier to any later point in the knitting. Popular motifs made with dip stitches include flowers, flower-stem leaves and fans.

ELONGATED STITCH

In knitting, an elongated stitch is a stitch that is longer than others. It may be created by wrapping the yarn around the righthand needle two or more times, or by placing yarn overs between stitches and dropping them on the next row. Elongation is often used in combination with other stitch techniques that produce textured effects in the fabric, as well as in lace work where it is used to create open areas. Works that involve elongated stitch knitting generally require blocking in order to finish the piece.

INCREASE (KNITTING)

In knitting, an increase is the creation of one or more new stitches, which may be done by various methods that create distinctive effects in the fabric.

METHODS OF SINGLE INCREASING (KNITTING)

- Yarn-over increase or "eyelet increase" — The simplest increase is to do a yarn-over between two existing stitches. On subsequent rows, the yarn-over will be knitted, making a new stitch. This disadvantage of this method is that a small hole (eyelet) is produced at the yarn-over. This can be improved by twisting the yarn-over stitch - similar to a "make one" (below).
- Raised Increase — Lift the strand connecting two knitted stitches in the row below onto the left needle (effectively producing a yarn-over) and knit it, either normally or twisted. This method (especially if twisted) leaves almost no hole, since forming the yarn-over stitch from the (presumably tight) connecting strand draws the two neighboring stitches together.

- Bar increase — Knit the stitch normally but without transferring the knitted stitch to the right needle; the same stitch is then knitted through the back loop. (Knitting through the front loop again is not feasible, since it would undo the first stitch.) This increase makes a bar or a nub on the fabric.
- Moss increase — Knit the stitch normally but without transferring the knitted stitch to the right needle; the same stitch is then purled. This increase makes a bar or nub on the fabric.
- Lifted Increase — For a right-side increase, knit into the right leg of the stitch of the row below the next stitch to be knit, then knit the next stitch. For a left-side increase, knit one stitch, then knit into the left leg of the stitch of the row below it. This kind of increase can be visually subtle.
- "Make One" (M1) — Place a half-hitch loop on the needle between two stitches, either before or after, and twisted either left or right, depending on desired effect.
- Column of Increases — A second strand of yarn or roving is passed up the piece. The second strand is used to make an extra stitch in each row by knitting a doubled stitch up from the lower row. The resulting piece has one more stitch in each row and each row is in the shape of an inverted V. The column of increases is used to make square sweater yokes.

Finally, a large number of increases in a row is best done by casting on; examples include buttonholes, etc.

ROLE IN FABRIC SHAPING

The typical use for increases and decreases is to taper a flat piece of knitted fabric, *e.g.*, in widening the sleeve of a sweater from the wrist to the biceps. However, increases and decreases can also be used to produce curvature, *i.e.*, to make the knitted fabric cup (positive curvature) or gather (negative curvature); in woven fabrics, this curvature is produced with darts, flares and gussets. A great advantage of knitted fabrics is that they can be shaped nearly invisibly without seams or cutting (as woven fabrics must) with increases, decreases, and/or short rows.

PLAITED STITCH (KNITTING)

In knitting, a plaited stitch is a single knitted stitch that is twisted clockwise (right over left) or counterclockwise (left over right), usually by one half-turn (180°) but sometimes by a full turn (360°) or more.Plaited stitches can be produced in several ways. Knitting into the back loop produces a clockwise plaited stitch in the lower stitch being knitted (*i.e.*, the loop that was on the left-hand needle.) The clockwise-plaited stitch is also called a left crossed stitch, since the left strand (*i.e.*, the outgoing strand) of the loop crosses over the right incoming strand. Left-crossed stitches are sometimes called twisted stitches, although the latter term might be confused with similar terms from cable knitting.

Conversely, a counterclockwise plaited stitch can be produced if the yarn is wrapped around the needle in the opposite direction as normal while knitting a stitch. Such a stitch is also called a "right crossed stitch", since the right incoming strand crosses over the left outgoing strand. Here, the plait appears in the upper stitch being knitted, *i.e.*, in the new loop being formed. In the "brute-force" approach, the knitter can produce any sort of plaiting by removing the stitch to be knitted from the left-hand needle, twisting it as desired, then returning it to the left-hand needle and knitting it.

Both clockwise and counterclockwise plaited stitches are often repeated in wales, *i.e.*, in columns of knitting, where they make attractive, subtly different ribbings. Fabrics made with plaited stitches are stiffer than normal and "draw in" sideways, *i.e.*, have a smaller widthwise gauge.

Extra-long, full-turn clockwise plaited stitches can be made by knitting through the back loop and wrapping the yarn twice; this is an attractive stitch when repeated in a row, creating openness and a change in scale that enlivens even simple stockinette or garter stitch. Plaited stitches are also useful in increases and decreases, both for drawing the fabric together and for covering potential "holes" in the fabric. As an aside, knitting through the back loop is a useful technique for untwisting stitches on the left-hand needle that "hang backwards". Such stitches are often produced when a knitted fabric is partially pulled out and some stitches are accidentally put back onto the needle with a backwards twist.

YARN OVER

In knitting, a yarn over is technique in which the yarn is passed over the right-hand knitting needle. In general, the new loop is knitted on the next row, either by itself (producing a hole) or together with an adjacent stitch (*e.g.*, in "tucked" slip stitches). The yarn-over may also be dropped on the next row, producing a longer stretch of yarn between the stitches of the previous row. Conversely, the effect of a yarn-over can be obtained by picking up the yarn between stitches of the previous row; the difference is that the yarn then is shorter, and the flanking stitches of the previous row may be overly drawn together.

The term "yarn-over" refers only to the act of wrapping the yarn around the needle, and not to the working of the next existing stitch. Yarn-overs are often used to increase the number of stitches, since knitting a yarn-over creates a new stitch where none existed previously, but does not use up a stitch on the needle. Yarn-overs are also common in eyelet and lace knitting, since they produce stable holes in the fabric.

Yarn-overs are also used to slip stitches neatly without having to pass the yarn in front or back. Instead, a yarn-over is done adjacent to the slipped stitch, and the two are knit together on the following row. Thus, the yarn is "tucked away" by passing *over* the slipped stitch, rather than in front or back. This is the basis for brioche knitting.

There are several types of yarn-over, depending on how many times the yarn is wrapped around the knitting needle and on the direction (chirality) with which the yarn is wrapped. Normally, the yarn is wrapped with a *right-handed* chirality, *i.e.*, counterclockwise when looking at the right-hand needle point-on. Wrapping the yarn the other way (*i.e.*, *left-handedly*) will result in a plaited stitch if the stitch is knit on the following row.

Wrapping the yarn multiple times produces very long loop, which can yield interesting visual effects. For example, a row of such long stitches makes an interesting horizontal stripe. Alternatively, the knitting of the long loop can be deferred until a later row (*e.g.*, 5 rows later); in this case, the loop need not be knit vertically, but can be knit diagonally, *e.g.*, to form the lower leaves of a flower.

4

Synthetic Fibres

Synthetic fibres are no substitute for structural (primary) reinforcement in concrete because they add little or no strength. But structural reinforcement doesn't provide its benefits until concrete hardens. That's why some contractors add synthetic fibres to concrete as secondary reinforcement. Unlike structural reinforcement, synthetic fibres provide benefits while concrete is still plastic. They also enhance some of the properties of hardened concrete. Synthetic fibres are most commonly added to concrete for slab-on-grade construction to reduce early plastic shrinkage cracking and increase impact- and abrasion- resistance and toughness.

The fibres also can be added to precast concrete to improve resistance to handling stresses, to pumped concrete to improve cohesiveness, and to shotcrete to reduce rebound and material waste.

HOW FIBRES CONTROL SHRINKAGE CRACKING

Shrinkage cracks are short, irregular cracks that can develop in concrete within the first 24 hours after concrete placement. Not to be confused with surface crazing, shrinkage cracks usually pass through the entire depth of the slab. The cracks are generally caused by either plastic or drying shrinkage. Plastic shrinkage occurs before concrete reaches initial set; drying shrinkage occurs after concrete sets. All concrete shrinks after placement due to a volume change caused by moisture loss. If the shrinkage could take place without any restraint, the concrete would not crack. But slab-on-grade concrete is always subject to at least some restraint by either the foundation, another part of the structure, or by reinforcing steel embedded in the concrete. Restraint also develops during differential shrinkage, when the concrete at the surface shrinks faster than the underlying concrete.

If the tensile stresses caused by restraint exceed the tensile strength of the concrete, the concrete cracks. Taking precautions can minimize the possibility of shrinkage cracking. These include keeping the concrete moist during curing and protecting the concrete with temporary coverings or curing compounds to reduce moisture evaporation. But even when you take precautions, shrinkage cracks can occur. Manufacturers of synthetic fibres say their products can be used to minimize plastic and early drying shrinkage cracking. Mixing the fibres in the concrete at the recommended dosages (usually 11D2 pounds per cubic yard) results in millions of fibres dispersed evenly throughout the concrete matrix. This multidimensional reinforcement reportedly gives fresh concrete more tensile capacity to resist typical volume changes. It also helps distribute tensile stresses more evenly.

If shrinkage cracks do form, fibres bridge these cracks, helping reduce their length and width. Various tests have been conducted on synthetic fibres to determine the effects of the fibres on plastic shrinkage cracking. Paul Kraai, a professor at

San Jose University, compared early cracking behaviour of slab samples made with plain concrete, concrete reinforced with welded-wire fabric, and concrete containing 3D 4-inch fibrillated polypropylene fibres at a rate of 11D2 pounds per cubic yard.

The slabs were 2 feet wide, 3 feet long, and 2 inches thick with edges restrained by panels. Fans blew air at high velocities over the surfaces of the slabs to force rapid drying. The results showed plastic shrinkage crack reductions of 71.5 per cent in the fibre- reinforced sample compared with the plain concrete sample. The sample containing welded-wire fabric showed crack reductions of only 6.5 per cent.

OTHER WAYS FIBRES ENHANCE CONCRETE PERFORMANCE

The effects of fibres on the behaviour of plastic and hardened concrete varies depending on the concrete materials, mix proportions, fibre type and length, and quantity of fibre added. Research from various sources, though, generally agrees that adding synthetic fibres to concrete can improve the following properties:

- Impact resistance (increases of 10 per cent to 50 per cent)
- Abrasion resistance (increases of 20 per cent to 52 per cent)
- Permeability (decreases of 33 per cent to 45 per cent)
- Toughness, or post-crack integrity (increases of 15 per cent)

Most of the research also shows, however, that the fibres provide only slight if any improvement to concrete compressive, tensile, and flexural strengths; modulus of elasticity; and splitting tensile strength.

Recognizing the performance characteristics of synthetic fibre-reinforced concrete, ASTM has established a standard test method for flexural toughness and first-crack strength of fibre- reinforced concrete (ASTM C 1018) and a standard specification for fibre-reinforced concrete and shotcrete (ASTM C 1116).

SYNTHETIC FIBRE TYPES

The number of synthetic fibre suppliers has grown in recent years, giving contractors a wide range of fibre products from which to choose. The primary types of synthetic fibres commercially available in the United States are polypropylene, polyester, and nylon.

Though the fibres within each type come in various lengths, thicknesses, and geometries, synthetic fibres provide similar benefits when used as secondary concrete reinforcement.

Table Typical Properties of Synthetic Fibers

Fiber	Type	Specific gravity	Length (inches)	Tensile strength (psi)	Young's modulus (ksi)
Polypropylene	Fibrillated	0.91	¼-2½ (or graded)	80-110	500-700
Polypropylene	Monofilament	0.91	½-¾	40-100	500-700
Polyester	Monofilament	1.34	¾-2	80-170	1450-2500
Nylon	Monofilament	1.16	¾-2	130	750

Fig. Fibrillated polypropylene fibres (left) consist of interconnected fibre strands that open during mixing (night)

They also are comparable in cost (about 1.5¢ to 2.5¢ per square foot per inch of concrete depth). Table 1 shows typical properties of the various synthetic fibres. Most synthetic fibres also have excellent acid, alkali, mildew, and salt resistance, are noncorrosive, and have low thermal conductivity.

POLYPROPYLENE

Of the synthetic fibres available in the United States, polypropylene is the most widely used in ready mixed concrete. Polypropylene fibres are hydrophobic, so they don't absorb water and have no effect on concrete mixing water requirements. They come as either fibrillated bundles or monofilaments. To produce fibrillated fibres, manufacturers extrude the polypropylene in sheets that are stretched and slit.

The result is a mesh of interconnected fibre strands rectangular in cross section. Manufacturers cut the strands to specified lengths and separate them into bundles. Fibre lengths range from 1D4 to 21D2 inches. When added to concrete during mixing, the fibrillated fibres open into a network of linked fibre filaments that mechanically anchor to the cement paste.

One manufacturer offers a graded fibrillated fibre having various lengths, sizes, and fibrillation patterns. The graded fibres reportedly disperse more thoroughly into all areas of the cement paste during mixing. Monofilament fibres are fine, cylindrical strands that separate during mixing. Because monofilament fibres are smooth and have a small surface area, they don't anchor into the cement matrix as well as fibrillated fibres. With fibrillated fibres, cement paste penetrates into the network of fibre filaments resulting in better mechanical anchoring to the concrete.

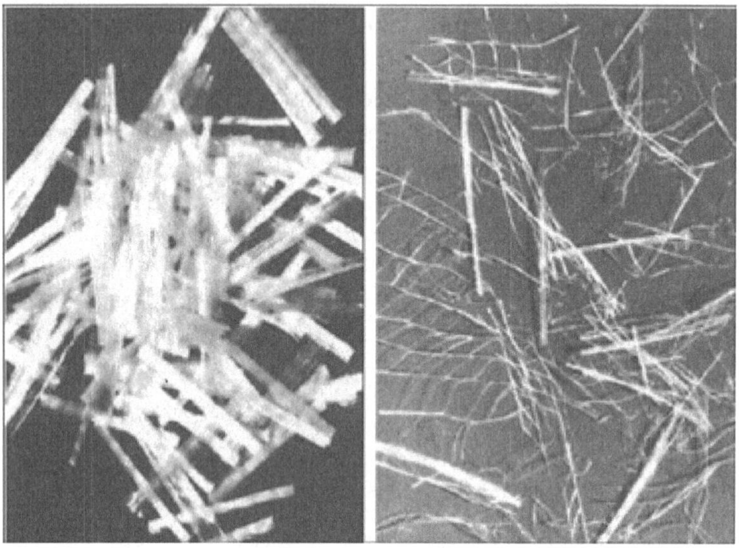

Research shows that lower volumes of fibrillated fibres than of monofilament fibres are needed to improve the post-cracking loadcarrying capacity and ductility of concrete. Some manufacturers recommend using monofilament fibres for relatively short-term benefits, such as plastic shrinkage crack control during the first few hours after concrete placement.

POLYESTER

Though not as widely used as polypropylene fibres, polyester fibres are offered by several manufacturers. The fibre bundles come only in monofilament form in lengths from 3D 4 to 2 inches. Like polypropylene, polyester fibres are hydrophobic. However, they have a tendency to disintegrate in the alkaline environment of portland cement concrete. To retard this degradation, manufacturers of polyester fibres coat the fibres to resist alkali attack. But the long-term performance of the coated fibres has not been determined.

NYLON

Like poly-ester fibres, nylon fibres come only in monofilament form. What primarily distinguishes them from polypropylene and polyester fibres is their hydrophilic nature. They retain a natural moisture balance of 4.5 per cent. Because of this strong affinity to water, nylon fibres bond chemically to the concrete matrix. The bond of polypropylene and polyester fibres is only mechanical. Nylon fibre manufacturers also report that their fibres have higher aspect ratios (ratio of length to diameter) than those made of polypropylene.

Therefore, they can be added in smaller dosages to produce the same reinforcing effects. Usually no more than 1 pound per cubic yard is needed.

TIPS ON USING SYNTHETIC FIBRES

Using synthetic fibres in concrete may require slight modifications to normal concreting procedures. Some fibre manufacturers recommend the following procedures for mixing, placing, and finishing fibre-reinforced concrete.

MIXING

Fibres can be added with the coarse and fine aggregate at the batch plant or to the central or truck mixer at the jobsite. If adding the fibres with other mix ingredients, no extra mixing time is needed. If adding the fibres to mixed concrete, agitate the concrete an additional 3 to 7 minutes as recommended by ASTM C 94 to disperse the fibres thoroughly. Because synthetic fibres don't affect the chemical hydration of cement, they work with all concrete mixes and admixtures without changing required mix proportions. To eliminate measuring, most fibre suppliers package the fibres in the recommended dosage for a cubic yard quantity of concrete.

Some fibre manufacturers package the fibres in cellulose bags that can be added unopened to the mixing drum. These bags disintegrate quickly in concrete, dispersing their contents. The disintegrated bag material has no effect on concrete properties. Ron Zollo, director of the International Fibrous Concrete Institute, says that uniform distribution of fibres was sometimes a problem years ago because the fibres would clump and form fibre balls in the mix. Manufacturers have eliminated this problem, though, by creating new fibre geometries that prevent fibre balls from forming.

PLACING

Because synthetic fibres mix into the cement paste, they don't interfere with structural reinforcement systems and they conform to unusually shaped forms, such as those used for curved driveways or some precast products. However, synthetic fibres can increase the stiffness and cohesiveness of the concrete mix. Workers may notice a slight decrease in workability when placing fibre-reinforced concrete. They also may notice a slump loss of up to 1 inch. The extent to which fibres affect slump depends on fibre design, length, and concentration.

Fibres only affect slump by increasing mix cohesion. They do not affect concrete's water- cement ratio. If the mix is too difficult to place, workers should not add extra water, which can weaken the hardened concrete. Instead, a superplasticizer should be used to improve workability. One polypropylene fibre manufacturer even prepackages its fibres with a superplasticizer in water-soluble bags. The superplasticizer reportedly maintains specified slump or increases slump up to 11D2 inches.

FINISHING

Synthetic fibres are compatible with all concrete surface treatments and finishes, such as pattern stamping, exposed aggregate, brooming, and hand or power troweling. The neutral fibres also won't stain or discolour concrete. However, a fibre-reinforced concrete slab usually bleeds less and more slowly than a plain slab because the fibres hold the concrete together and reduce segregation. So it's especially important when finishing fibre-reinforced concrete not to start too early. Make sure all bleedwater is gone. It's even wise to start floating and troweling closer to final set than normal to help avoid fibre exposure on the slab surface.

If any time during finishing, you notice the fibres being pulled up, you may be on the concrete too soon. When finishing fibre-reinforced concrete, use only steel or magnesium trowels and floats. Wood trowels and tools are more abrasive to the surface and may reveal more fibres. Also, don't use tined garden rakes to move or place the concrete. The tines can disrupt fibre distribution. For textured or broom finishes, use a broom with stiff bristles and broom in only one direction.

This will cause the fibres at or near the concrete surface to align within the ridges so they are not apparent If mixed into concrete thoroughly, fibres should not float to the concrete surface nor sink to the bottom. Still, fibres may appear on or near the surface because they are distributed throughout the concrete. Any surface fibres that do appear will wear off quickly under light traffic. If a surface coating is specified, burn off exposed fibres with a propane torch before applying the coating.

FIBRES VS. WIRE MESH

Both synthetic fibres and wire mesh used as secondary reinforcement can help control cracking in cast-in-place and precast concrete. The primary differences are when and how they work. Fibres are most beneficial soon after concrete placement by controlling the formation of plastic shrinkage cracks. Wire mesh, on the other hand, does not prevent crack formation. Instead, the mesh holds cracks together *after* they have formed. From an economic standpoint, using fibres eliminates the costs of handling and placing wire mesh. Also, wire mesh must be placed properly to be effective.

This is not a concern with fibres, which disperse evenly throughout the concrete during mixing. And unlike wire mesh, synthetic fibres are noncorrosive and won't rust. An ideal secondary reinforcement system would use both fibres and properly placed wire mesh. But this approach is not always practical or cost-effective. If you must choose between fibres and wire mesh, base the decision on the benefits and limitations of each product.

FIBRES IN SHOTCRETE AND PRECAST CONCRETE

Though synthetic fibres are most commonly used in cast-inplace concrete, their use in shotcrete and precast products is growing. Adding synthetic fibres to shotcrete can reduce rebound and material waste, permit thicker layers per pass, reduce sagging, and inhibit plastic shrinkage cracking. Shotcrete reinforced with a high volume of fibrillated polypropylene fibres was first used successfully in 1988 for a tunnel lining project in Alberta, Canada. The shotcrete contained 11D2-inch-long fibres at a minimum content of 10.1 pounds per cubic yard.

Even with these high fibre volumes, the shotcrete was easy to mix with standard equipment. It also pumped easily and there was no rebound during application Precasters use synthetic fibres in their products to add early strength, help reduce concrete breakage during handling, and save on labour. They also report fewer shrinkage cracks when using the fibres. Some precasters use synthetic fibres in unusually shaped precast products, such as planters and urns, instead of wire reinforcing. If not carefully placed, wire reinforcing in these products is sometimes exposed when the products are removed from the forms.

CHEMICAL AND PHYSICAL PROPERTIES OF SYNTHETIC FIBRES

The oceans, seas and other water cover more than 70 per cent of the earth's surface. They occupy about 90 million sq. miles and yield approximately 80 million tons of fish annually. Fishing is one of the oldest professions of man. Fisheries have developed continuously over the centuries, utilising improved and larger ships, more sophisticated fishing equipment and catch preservation techniques. Modern fishing has developed through three main technological revolutions.

Mechanisation begins in the second half of the nineteenth century with the use of steam driven capstans. Steam power was successfully used in towing a trawl for the first time in 1877 and from that time steam gradually replaced manual and sail propulsion before being superseded by the internal combustion engine. Today, the engine is universally favoured for larger vessels while petrol driven outboards remain common for small boats.

Introduction of electronic fish finding equipment started half a century ago. Aircraft enable large areas to be searched visually, while satellite and laser technology are being utilised as aids to identify suitable environmental conditions and in finding fish schools. Recent developments in microprocessor technology enable rapid analysis of signals from high performance transducers. Today microprocessor technology also permits electronic navigation, fish detection, ship handling and fishing gear control to be integrated and fishing operations to be increasingly automated. The advent of synthetic fibres heralded the third major revolution in modern fishing gear. Until recently, natural fibres such as sisal, manila and cotton were by far the most important raw materials for ropes and nets, and only during the last two or three decades have they been gradually replaced by synthetic fibres. Synthetic fibres are man-made from fibres, which have been produced entirely by chemical synthesis from simple basic substances.

As compared with vegetable fibres they are of better uniformity and continuity, have higher breaking strength and are more resistant to rotting. These materials have greatly extended the endurance of gear, and together with the mechanised net and rope making have increased the size and complexity of nets. The development of synthetic fibres started around 1920 by Staudinger, winner of the Nobel Prize in chemistry in 1953. Staudinger found that all fibrous material consists of long chain molecules in which a great number of equal, simple units are linked together.

This structure gives the fibrous material the properties required from a textile fibre. Based on this knowledge, a great deal of other chemical research has been carried out in recent years to create such fibre-forming macromolecules. At the present many groups of synthetics fibres are produced for the fishing industry. The most important countries manufacturing manmade fibres are USA, Japan, Netherlands, Great Britain and France. Developments of new synthetic materials and fabrication techniques play an important role in improving fishing gear efficiency.

However, as different classes of synthetic fibres become available, the fishermen's and the netmaker's knowledge of the different materials and their characteristics is often not good enough to allow optimum selection and utilisation (principally in the developing countries). The objective of this project is to compare the general characteristics of the synthetic fibres most commonly used in fishing gear (polyamide, polyester, polyethylene, polypropylene, polyvinyl and dyneema) and to provide essential information on their properties, with reference to their uses and application in Cape Verde fisheries. Dyneema is the newest of these fibres and will be given special attention in this study.

CAPE VERDE ISLANDS AND THEIR FISHERIES

The Archipelago of Cape Verde is located 450 km off the coast of Senegal, West Africa, and is composed of 10 islands and 8 islets. The islands are of volcanic origin and positioned between 15° and 17° N. The islands are divided into two groups named according to the trade winds that reach them from the African continent: windward and leeward. The first group consists of the islands of Santo Antao, Santa Luzia, São Nicolau, Sal and Boavista. The second group more to the south, comprises the islands of Maio, Santiago, Fogo and Brava. With a total of 4,033 km2, Cape Verde is one of the five Atlantic archipelagos of the Macronesia.

The others are Azores, Madeira, The Canary Islands and Savage Isles. The fisheries sector contributes about 5 per cent to the GNP of Cape Verde. It is important for the local economy and employs over 7000 fishermen and 3200 fish traders. Fish is an important source of nutrition to the population and constitutes 70-90 per cent of animal protein in the diet. Consumption of fish and fisheries products is 18,5 kg/year per capita.

The fisheries potential in Cape Verde waters is estimated at around 30.000 - 35.000 tons but only 26 per cent of this potential is exploited. The total catch in 1997 was 9160 tons. In general the fisheries can be divided into industrial and artisanal fisheries.

The industrial fleet (>11 m) is composed of 32 small vessels <20 tons GRT each, mainly catching pelagic species such as tuna and horse mackerel. The artisanal fleet is composed of 1455 small vessels locally called "botes" (towing boats with outboard engines), catching tuna, small pelagic and demersal species. Industrial fishermen use mainly hand-lines, purse seines, and traps and artisanal fishermen use hand-lines, small purse seines and beach seines. Most fishermen in Cape Verde, lack formal training.

The knowledge they possess is gained from experience or learned from their fathers. So fishing gear is constructed in an empirical manner without consideration for the quality of the materials to be used in the construction process. The principal synthetic materials used in Capverdian fishing gear are listed in Table.

Table. Principal syncipal synthetic materials use in Capverdian fishing gear

	Materials	
Gear	Type	Specification
Purse seine (small pelagic and mackerel)	Netting	PA R 155 tex 24mm, PA R 230 tex 24 mm, PA R 310 tex 24 mm, PA R 310tex 20 mm, PA R 860 tex 120 mm, PA R 310 tex 16 mm, PA R 230 tex 16 mm, PA R 620 tex 100 mm
	Ropes	PA 8, 10, 16 mm, PE 16 mm
Beach seine	Netting	PA R 310 tex 50 mm, PA R 310 tex 30 mm, PA R 310 tex 24 mm, PA R 390 tex 20 mm, PA R 390 tex 16 mm
	Ropes	PA 10, 12mm, PE 14 mm
Gill net	Netting	PA mono 60 mm ø 0.30 mm, PA 160 mm R 950 tex, PA 60 mm R 230 tex PA 6, 10, 16 mm, PE 8 mm, PP 3
Traps	Ropes	PES 22 mm

Most of these materials are imported from Portugal by the National Institute of Fishing Development (INDP) which stipulates the price. INDP has commercial centres distributed in fishing communities all over the country. The Institute doesn't pay tax on imported materials and therefore it is difficult for private companies to compete with them. In addition, the materials are limited and their quality is variable.

To keep up with developments, new knowledge in fishing industry is important for Cape Verde. The better Capverdian fishermen understand the properties of the various fibres, the better they will be able to employ them in the most useful and appropriate way. The knowledge gained in this study will be used in the training of fishermen and netmakers with the objective of improving fishing gear in Cape Verde. This study will also provide information of interest to entrepreneurs.

CHARACTERISTICS OF SYNTHETIC MATERIALS

Chemical Classification of Synthetic Fibres

The chemical groups or classes of synthetic fibres that are used in fishing gear are listed in Table.

Table. Chemical groups or classes of synthetic fibres.

Group	Symbol	Some trade names
Polyamide	PA	Nylon, Amilan, Anzalon, Enkalon, Kapron, Perlon
Polyester	PES	Terylene, Dacron, Diolen, tergal, Tenital, terlenka, Tetoron
Polyethylene	PE	Corfiplaste, Courlene, Drylene, Etylon, Kanelight, Nymplex, Polythene
Polypropylene	PP	Meraclon, Courlene PY, Danaflex, Hostalen P, Nufil, Ulstron
Polyvinyl Alcohol	PVAA	Cremona, kuralon, Kuremona, Manryo, Vinylon.
Dyneema	SK 60, SK 75	Dyneema SK 60, Dyeema SK 75, Dynex

Polyamide (PA) - hvort er PA 66 eða 6.6?

Several types of PA fibres are manufactured, differing in their chemical components. The most important types are PA 66 and PA 6. PA 66 is manufactured from two components, hexamethylenediamine and adipic acid, each containing six carbon atoms. The PA 6.6 was discovered in 1933 by W.H. Carothers in the USA and is called nylon.

PA 6 is made from one monomer called caprolactam, which contains six carbon atoms and was first synthesised in 1938 by P. Slack. With regard to the fisheries there is no difference between PA 6.6 and PA 6. They are both manufactured in the same manner, and are equally suitable for fishing gear. Therefore, no distinction will be made between the two PA types when discussing the properties of netting yarn in this paper.

Polyester (PES)

The principal PES fibres are made from polyethylene terepthalate, and were first synthesised by Whinfield and Dickson in the laboratories of the Calico Printers Association in Great Britain. Mr. Whinfield named the fibre "Terylene" .Polyethylene (PE)

PE fibres, used for fishing gear, are produced by a method developed by a German, Ziegler, in the early 1950s. The monomer ethylene, the basic substance of polyethylene, is normally obtained by cracking petroleum. The same applies to propylene, the basic substance for producing polypropylene.

Polypropylene (PP)

Natta, winner of the Nobel Prize in chemistry, developed the synthesis of polypropylene in 1954. PP and PE are often collectively called "polyolefines". Here they are seperated into two groups because of their different properties with regard to fishing nets.

Polyvinyl Alcohol (PVA)

PVA fibres, whose production is based on the research of Hermann and Haehnel, have been greatly improved in Japan since 1938. It was also the first synthetic material to be used for fishing gear and the first to demonstrate the immense practical advantages of rot-free fibres. Today their use is very limited in fishing gears, so these fibres have not been included in the present study.

Dyneema

Dyneema fibres are produced by gel spinning, a process invented and patented by the Du Point company (DSM) in 1979 and they have been in commercial production since 1990 at a plant in the Netherlands. " Polyethylene with an ultra high molecular weight (UHMWPE) is used as the starting material. In normal polyethylene the molecules are not orientated and are easily torn apart.

In the gel spinning process the molecules are dissolved in a solvent and spun through a spinneret. In the solution the molecules that form clusters in the solid state become disentangled and remain in that state after the solution is cooled to give filaments. As the fibre is drawn, a very high level of macromolecular orientation is attained resulting in a fibre with a very high tenacity and modulus ". Dyneema fibres are produced in the following commercial grades:

- Dyneema Sk60 / Sk75 are the multi-purpose grades. These are used for ropes, fishing nets, cordage and protective clothes.
- Dyneema Sk 75 has a higher tenacity than Dyneema Sk 60. This grade is specially designed for ropes, cordage, fisheries and textile applications.

Identification of Synthetic Materials

Various kinds of synthetic fibres are similar in appearance. However simple methods can be used to identify synthetic materials. The methods most commonly used by fishermen are:

- The water test, which classifies the netting material according to whether it floats or sinks,
- Burning. The material is identified from the smell of the smoke and the residue.

MATERIAL AND METHODS

Material

In the implementation of this project, tests on physical and chemical properties of PA, PE, PES, PP, Dyneema Sk 60 and Dyneema Sk 75 were to indentify the characteristics of each of them. The following properties, relevant for their use in the fishing industry, were examined: Breaking strength, elongation, elasticity, resistance to abrasion, strength, density and the changes of length in water.

The tests were conducted from 25 October to 17 November 1999 at Hampidjan, which is a leading manufacturer of high-quality trawls, trawl netting, ropes and others products for deep-sea fisheries. This company has a long history in fishing gear production, dating back to its foundation in 1934. The samples were collected from materials readily available in the Icelandic market. In the tests for breaking strength, elongation, elasticity and resistance to abrasion, 14 mm, three strand ropes were used. Dyneema ropes are not available in three strand so only the diameter was considered. The choice of the diameter was based on its popularity among fishermen, availability in the market and the capacity of the available equipment in Hampidjan laboratories. Some characteristics of the material used in the tests are show in Table.

Table : Material used in the tests for breaking strength, elongation, elasticity, and resistance to abrasion (g/m⇒ gram/meter, denier⇒ Weight of 9000 m).

	Materials				
Characteristics	PA	PE	PES	PP	Dyneema
Weight g/m	104.5	97.8	157	81.2	137.5
Weight denier	938	880	1413	731	1237.5

To measure length changes in water, 1, 1.5, 2.0, 2.5, 3.0, 3.5, 4 and 4.5 mm netting was used. Methods recommended by ISO (International Organisation of Standardisation) were used in most of the tests. The standard temperature and atmospheric pressure for testing is specified for physical tests of textile materials in dry conditions. This requires a relative humidity of 65 per cent and a temperature of 20°C. The Hampidjan laboratory complies with the temperature requirements. Humidity was not measured but variation in humidity in Iceland is small and did not cause any problems during the testing period. All tests were repeated ten times. The results were recorded on data sheets.

Tests for Breaking Strength

Breaking strength is the maximum load (force) needed for the material to break. Breaking strength was determined by using a special tensile testing machine made by Koyo Ltd - Japan. It is equipped with one electronic force meter.

This machine can apply force up to 20 tons. The sample is fastened to the clamp of the machine and extended under increasing force until the sample breaks. The electronic force meter registers the force in kilo Newtons needed to break the sample.

1 kilogram-force (kgf) = 9.81 Newtons (N)
1 deca Newton (daN) = 1.02 kilogram-force (kgf)

Fig. Tensile testing machine in the Hampidjan laboratory, used to measure

Free breaking length or strength

Free breaking length or strength is the theoretical length of a fibre, yarn or rope when it breaks under its own weight, when hanging freely. Amount of strength can be calculated by the formula:

$$\text{Strength (g/ den)} = \frac{\text{Breaking strength (Kg)}}{\text{Weight of material in denier}}$$

Tests for elongation

Elongation is the increase in length of a specimen during a tensile test. Due the limitation of time and facilities (equipment), only the elongation obtained immediately after the application of a specific tension was observed. A tensile testing machine was used to determine elongation. Samples of the same dimensions and characteristics as those used in the tests for breaking strength were fastened to the clamps of the machine and submitted to a pre-tension of 100 kN. The length between the two eye splices was measured and specimen was stressed to 75 per cent of its breaking strength and the length measured again. The length increase is the elongation. Elongation is expressed as per cent increase of the initial length of the specimen.

$$\text{Elongation(\%)} = \frac{\text{Increase in length}}{\text{Original length}} \times 100$$

Tests for Elasticity

Elasticity is the ability of material to return its original shape and dimension upon cessation of a deforming force. After finishing the tests for elongation, the force (75 per cent of the breaking strength) was released and the sample submitted again to a pre-tension of 100 kN. The distance between the two eye splices was measured and this measurement is called elastic elongation. The degree of elasticity is presented as the per cent decrease in total elongation and is calculated according to the formula:

$$\text{Degree of elasticity (\%)} = \frac{\text{Elastic elongation x100}}{\text{Total elongation}}$$

Tests for Abrasion Resistance

Abrasion resistance is the resistance of a material or specimen against wear when rubbed against a surface. This property is important for gear, which operates at the bottom and suffers wear and tear during towing. Quantitative measurements of wear and tear of fishing materials on board boats is very difficult. Therefore, the situation needs to be simulated in a laboratory. There are many methods of simulation but Hampidjan has developed an apparatus that approaches the "real situation" well. This apparatus consists of four pulleys of different sizes (two mechanically driven pulleys and two others, indirectly driven with the help of the first two). The pulleys are used to adjust the required abrasion speed and pressure on the specimen during testing and a small tap for frequently spraying the specimen with water.

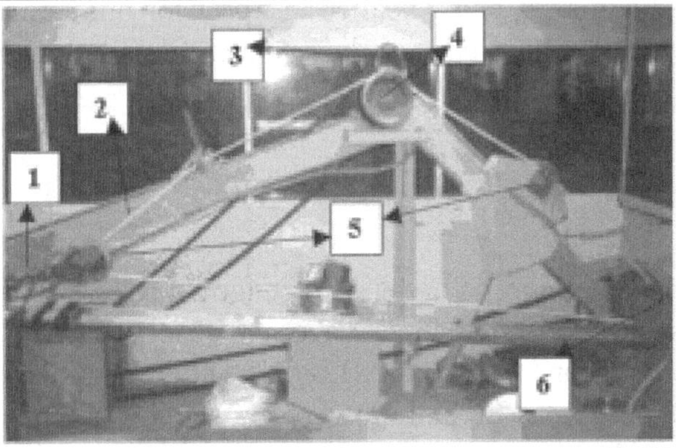

1. Apparatus to exert tension in the sample
2. Sample
3. Tap
4. Mechanically driven pulley
5. Simple pulleys
6. Mechanically driven pulley

Fig. Apparatus for testing abrasion resistance in the Hampidjan laboratories.

The sample is tested by putting it into contact with the abrasive surfaces of pulleys of the testing apparatus for 8 hours. Abrasion is measured as the loss of breaking strength. The results are expressed in per cent decrease of initial breaking strength.

Length Changes in Water

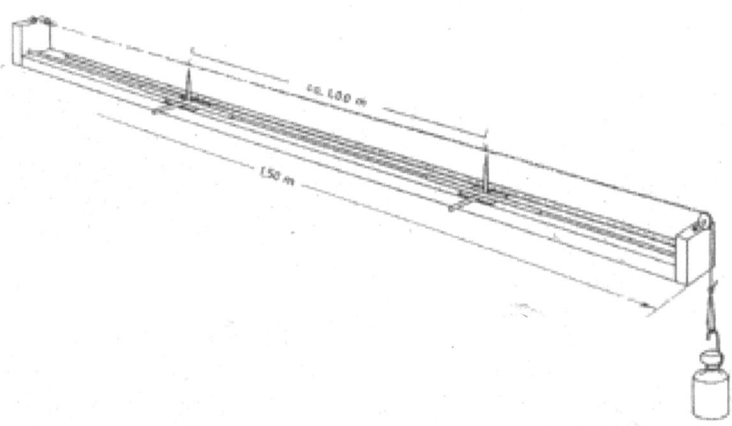

Fig. Apparatus for measuring the lenth of netting yarn (ISO 1974)

The length change in water was determined with an apparatus closely resembling the one illustrated in Fig and built with help of Mr. Larus Palmason, Director of fishing department in Sudurnesja Comprehensive College (Keflavik) and supervisor in this project. The tests were carried out on netting yarn, continuous filament 1.0, 1.5, 2.0, 2.5, 3.0, 3.5, 4 and 4.5 mm in diameter. According to ISO recommendation, the tests were conducted under dry conditions (standard atmosphere) and the length measurements were carried out under a pre-tension on the specimens corresponding to the mass of 250 m of the respective netting yarn. The netting yarn sample of about 100 cm was first measured dry, then again after being soaked in water for one hour. The results are expressed in per cent increase or decrease of the initial length of the specimens.

RESULTS

Breaking Strength

Figure shows the average recorded breaking strength of ropes, 14 mm in diameter, based on 10 observations. The lowest average observed was for PE (23.4 kN) and the highest was for Dyneema (221.3 kN). Difference in average breaking strength between the PA, PES and PP were not significant. Dyneema was significantly stronger than any of the other materials and PE significantly weaker.

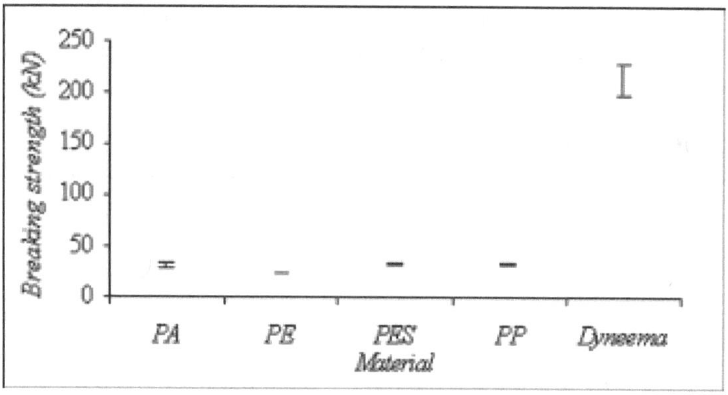

Fig. Average breaking strength of 14 mm ropes of different materials based on 10 tests.

Free Breaking Strength or Strength

The calculated values of free breaking strength, based on breaking strength and the density of material are presented in table. Dyneema is in a class of its own, but the order of the other materials is different from the one obtained in the tests of breaking strength.

Table: Calculated values of free breaking length for ropes with 14 mm of diameter

Material	PA	PE	PES	PP	Dyneema
Strength (g/den)	3.41	2.72	2.36	4.49	17.49

Elongation

The differences in elongation between PA, PE, PES, PP and Dyneema were highly significant ($F = 180.6$, $P < 0.05$, F crit $= 2.57$) and according to pair-wise t-tests, all of them were significantly different from one another. Dyneema changed its shape least under pressure while PA stretched the most. Fig: Elongation of 14 mm ropes. The bars are the standard deviation of the means. All the specimens strained by a load corresponding to 75 per cent of respective breaking strength.

Elasticity

The elasticity of 14 mm ropes of different kinds of synthetic material were tested in dry conditions is shown in figure. The values are presented as a per cent decrease in total elongation. Dyneema had the highest average elasticity and PP the lowest.

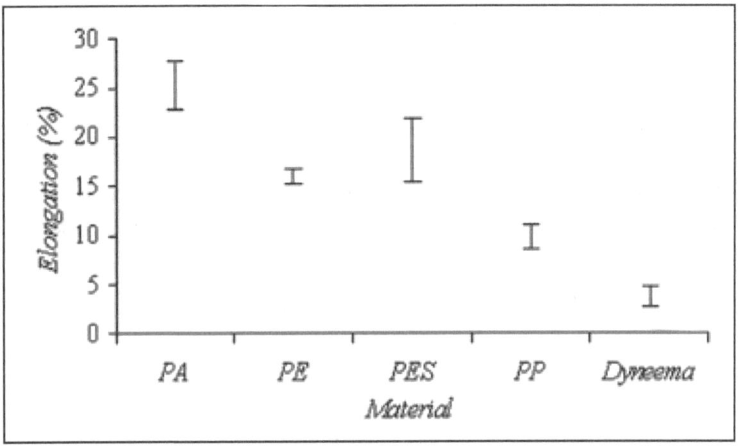

Fig. Elasticity of 14 mm ropes. The bars are the standard deviation of the means.

The difference in elasticity between PA, PE, PES, PP and Dyneema was significant (F = 3.20, P=0,02, F=2.58).

Resistance to Abrasion

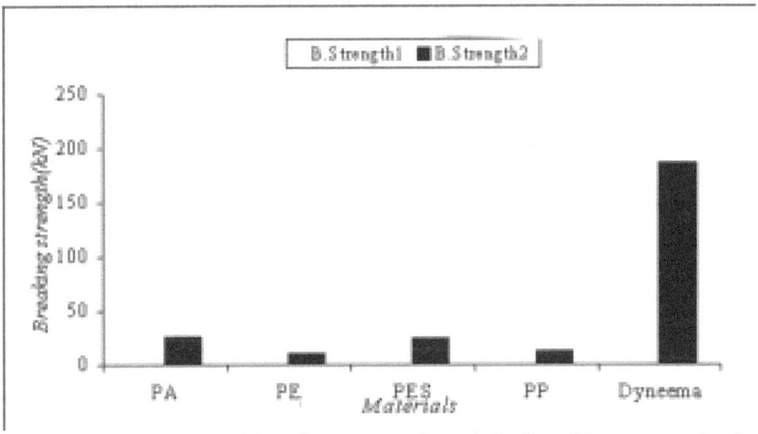

Fig. Comparison of original breaking strengh and the breaking strength after 8 hours of abrasion.

Shows the average initial breaking strength in comparison to the breaking strength after the material had been submitted to 8 hours of abrasion. All materials lost same strength during the tests. In relative terms, the loss of strength was least in Dyneema and most in PP.

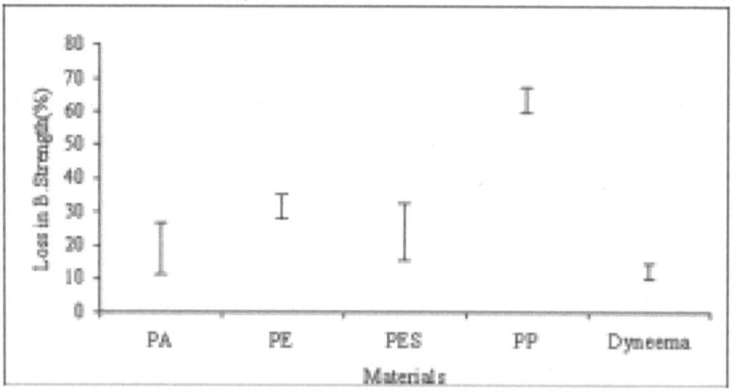

Fig. Loss in breaking strength after materials being submitting to abrasion for 8 hours.

ANOVA tests indicate that the difference in resistance to abrasion between PA; PE; PES, PP and Dyneema were highly significant (F = 86.4, F crit = 2.63) as was the difference between PA, PE. PES and PP (F=98.4, F crit = 2.88). The loss of breaking strength of PA was not significantly greater than in Dyneema (F=8.7, F crit =3.3).

Length changes in water

Netting yarn made of PES, PE, PP and Dyneema fibres remained practically unaffected by water and it had no significant influence on length. PA netting yarns react differently in water. Some shrink, some lengthen and others remain unchanged. The changes in length of netting yarn 1, 1.5, 2.0, 2.5, 3.0, 3.5, 4, 4.5 mm in diameter whilst in water are shown in table.

Table. Changes in length for netting yarn in water. Pre-tension 0.25 gf.tex

Changes in length (W)										
Materials										
Mm	PA		PE		PES		PP		Dyneema	
	Rtex	W (%)	Rtex	W (%)	Rtex	W (%)	Rtex	W (%)	Rtex	W (%)
0.5	241	-2								
1	764	-2	694	0	930	0				
1.5	1677	+2	1099	0	2116	0	1163	0.5	535	0
2	2931	0	2105	0	3734	0	2046	0		
2.5	4519	+3	3210	0			3232	0	2046	0
3	6437	+4	4717	0			4953	0	3232	0
3.5	8680	+3	6579	0						
4										

DISCUSSION

In the present study only 14 mm rope was tested, and caution should be taken in extrapolating the results to ropes of different diameters. Ultimately the best indicator of the quality of a rope is the breaking strength. The tests of breaking strength were done with two eye splices, one at each end. Eye splices at the end of the rope will cause approximately 5 per cent loss in breaking strength. In the research this loss was not taken in consideration so, the results underestimate the true breaking strength for ropes of 14 mm. According with the international standard 1805, all tensile testing machines shall include a pair of suitable devises for to hold the specimen. Due to some problems with the equipment only 3 tests were done on Dyneema. For other materials, tests were repeated 10 times. It is doubtful that the abrasion test adequately reflects the real situation. In spite of the shortcomings of the study listed above, there is enough information to allow a realistic comparison of the materials tested. The following observations can be made about the results of this study and constitute the main points of the results:

- Synthetic materials are generally strong. The breaking strength of Dyneema is greater than that of other material. This aspect is the most important quality of this synthetic fibre.
- According to the Dyneema manufacturer, the material has much higher abrasion resistance compared with other synthetic fibres. This statement should be taken with some reservation. In this study Dyneema lost 12 per cent of the initial breaking strength in an abrasion test, which was less than found for other materials. But this loss represented 92 kN which in absolute terms is a greater loss than was observed in any of the others materials. During the abrasion tests we observed that Dyneema rope in contact with unpolished or corroded steel showed more visible signs of damage than the other materials. Flexing and bending however was no problem and the twines remained flexible.
- PA netting yarns react differently in water. Some shrink, some lengthen and some remain unchanged. Unfortunately this study does not provide enough data to discuss the significance of these changes with regard to mesh size, but according to Klust the influence of water on the mesh size is in principle the same as on length of netting yarns.

Shows that the synthetic materials are all very good, although no ideal material exists. The problem therefore is to select the best available material for a specific purpose. The selection will depend on the quality of the material, the type of gear to be constructed, how and where it will be used and not least it will depend on price and availability. Bottom trawl nets require materials, which do not float, has high breaking strength, high elongation and elasticity, and high abrasion resistance. PA, PES and PE are at present the most commonly used materials in bottom trawls. Although not perfect, PA and PES meet most of criteria. PE however, is less expensive. Most of the properties required for bottom trawls are also required for mid-water trawls. PA is at present the most commonly used material for mid-water trawls but Dyneema is a good substitute for PA.

Table. Properties of ropes with 14 mm made of synthetic fibres.

Properties	PA	PE	PES	PP	Dyneema
Breaking strength (kN)	31 (V. good)	23 (Good)	31 (V. good)	32 (V. good)	212 (Exc.)
Strength (g/den)	3.41 (Good)	2.72 (Not good)	2.36 (Poor)	4.49 (V. good)	17.4 (Exc.)
Elongation (%)	25	16	19	10	4
	(Exc.)	Good	(V. good)	(Not good)	(Poor)
Elasticity (%)	19 (Good)	31 (Not good)	24 (Not good)	63 (Poor)	12 (V. good)
Res. to abrasion (% loss in breaking strength)	19 Good	31 Not good	24 Not good	63 Poor	12 V. good
Length changes in water	Yes	No	No	No	No
Density	Sinks	Floats	Sinks	Float	Neutral

V.Good = Very good; Exc.= excellent

The use of Dyneema twines in nets or parts of nets will substantially reduce the weight of the net. This will of course reduce the resistance in the water and the drag.

The lower drag can also be exploited by towing a larger net with the same engine capacity and so raising fishing efficiency. In general, fish have a greater chance of escaping a mid-water trawl than a bottom trawl. By making the mid-water trawls larger this disadvantage can be reduced to some extent. Hampidjan has recently constructed the world's largest mid-water trawl. This trawl, called Gloria, is made of Dyneema.

The main drawback of Dyneema is the high price. For purse seines the requirements are determined by the operational characteristics of the gear. The most important features are fast sinking rates, high breaking strength, low resistance to water and a good price. At present no materials have all these properties. On the whole the most suitable material is PA. Due to the low elongation, the capacity to float and the excellent elasticity Dyneema is maybe suitable for use in the float rope. Gillnets are set on the bottom, anchored floating, drifting or encircling.

Material for gillnets should have the lowest possible visibility in water, enough strength to withstand the forces of the entangled fish, appropriate elongation and elasticity. No materials combine all these properties, but due to their transparency PA monofilament are the most common. In gillnets, Dyneema is not a good choice. Apart from the prohibitive price, it has a low elongation, reducing the catcability of the gillnet. For a long-line all of synthetic fibres can be used.

Cape Verde fisheries are small-scale, the fishing gear is fairly basic, but even so, material represent the major part of the total fishing investment. Cape Verde Islands are of volcanic origin and the bottom is rough, causing a great deal of wear and tear of gear. The current on the bottom is strong and consequently the risk of gear loss is high. Fishermen have financial limitation and little knowledge of the properties of materials. For these reasons the material actually used in fishing gear will not depend so much on technical suitability as on local availability and price. It is important for Cape Verde fisheries that private companies compete with INDP. To make this possible the government needs to accord the same facilities and support to all entrepreneurs.

PA and PE are the most commonly used materials in Cape Verde fisheries. PA netting is used in purse seines, beach seines and gillnets (mono-filament) and PE in the work ropes (float ropes, footrope). PA is used correctly but the use of PE is not appropriate. The results of this study show that the physical properties of PE (lower breaking strength, less elasticity, poor abrasion resistance) compare unfavourably with those of PA, PES and PP. The popularity of PE is partly due to the low price. In Cape Verde PE is the cheapest of all synthetic fibre products.

Fish aggregating devices (FAD) are used in Cape Verde for concentrating fish in sufficient numbers for purse seining and to localise the fishing ground. FAD is a non-rigid structure which is secured to the bottom by an anchor, rises vertically through the water column and is suspend by means of a surface buoy. Devices designed to attract and temporarily retain desired fish species are attached to the vertical mooring line. FADs are moored both in shallow and deep water (between 150 - 2000 meters depth). The must common materials used in the anchor line are braided PP and PA. The diameter and the length depend on the depth of the water. Due the to the high wear and tear in Cape Verde and the strong currents the lifetime of FADs in deep waters is approximately 3 months.

Due to the importance of FADs and the cost it is recommended that INDP make some experiments using Dyneema Sk7 in the anchor line. Saga Petroleum has tested a Dyneema rope in the mooring of an oil rig in 280 m deep water at the Norwegian coast.

According to this study Dyneema is much lighter and about the same strength as a wire rope of same diameter. Compared to a polyester rope it is much lighter as well as much thinner. These are great advantages, but most important in handling is the fact that a Dyneema rope has neutral buoyancy. This means that the pressure on the rope is practically independent of its length. This results in a relatively simple installation procedure for the anchors. The use of Dyneema will increase the cost of the FADs but this might be offset by a longer lifetime of the FADs.

SYNTHETIC RESIN FIBRES

Nylon differs radically in its physical and chemical properties from other synthetic fibres. It resembles silk more closely than it does any of the other natural fibres. The great strength, elasticity, and wearing qualities claimed for it have aroused considerable interest.

The advantages of nylon are its high strength and elasticity, its lack of sensitivity towards moisture and various chemical agents, and its ability to be spun into fine filaments. It is stronger and more elastic than silk. It is nearly as strong when dry as when wet. Because it does not absorb moisture readily, garments made of it dry quickly. Mineral acids deteriorate it and phenol and formic acids are active solvents. No common dry cleaning solvents affect it, and it is quite resistant to alkalies even in concentrated form and also to mildew, molds and moths.

Nylon does not burn. When held in an open flame it reacts similarly to silk, wool or acetate rayon. The fibres melt and fuse into a glassy round-shaped mass. Varied commercial uses have been made of nylon, and it probably will replace silk to a great extent in the textile field. This fibre can be made into fine hosiery and other knit goods such as underwear. Some other uses are sewing thread, brush bristles, racquet and violin strings, surgical sutures, fishing lines, woven dress goods and velvets.

Vinyon is a synthetic fibre having remarkable properties. It is thermoplastic and when heated above 117° F. will shrink. If held in a flame it melts and chars but does not burn. Concentrated acids or alkalies will not injure vinyon fabrics. The fabric does not absorb moisture, consequently it will not mildew and will be quite useful in damp climates. This fibre has high elasticity and break ing strength; and because it is not affected by water, its strength is the same wet as when dry.

Vinyon has been put to many uses. Short fibres of vinyon have been mixed with cotton, wool, glass or other fibres to form felts. The most extensive use is for industrial filter cloth because of its high dry and wet breaking strength and its resistance to acids and alkalies. Some other uses are fish lines, nets, seines, acid and alkali-resistant clothing, electric insulation, shower curtains, bathing suits, waterproof clothing, fireproof awnings and curtains. Vinyon has also been used for shoe fabrics and gloves. Stains are readily removed from vinyon fabrics in a soap solution, and the material dries quickly.

GLASS FIBRES

Glass fibres are smaller than other textile fibres. Glass in its solid form has hardness, strength, cleanliness, great durability, and is plastic when molten. Glass fibre has all these desirable attributes of solid glass. The properties of glass fibre make it of value in many ways. It is a poor conductor of heat and electricity, and thus is used extensively for heat and electrical insulation. It can be subjected to temperatures of 1,000° F. without melting. It is inorganic and does not absorb moisture; consequently it resists decay, does not mildew, and vermin will not eat it. Glass fibres are very resistant to alkalies, acids and other chemicals. Damasks, brocades, satins, taffetas and sheer nets-all of varying textures, handles and sheens-are being made of glass.

These are for overdrapes, glass curtains, shower curtains, bedspreads, lampshades, awnings and other articles. Glass fiber is not suitable for dress fabrics. For wearing apparel it is used in shoes, hats, belts, neckties, handbags and other dress accessories. The new glass fabrics have advantages over fabrics of rayon or natural fibers because they are colour fast, durable, will not shrink, are resistant to heat, and are practically soil proof. Dust or dirt can be wiped off with a damp cloth. Table cloths and other articles when laundered should be flexed and rubbed as little as possible because the brittle fibers lose strength when abraded.

MODIFIED FIBERS

Fibers made by the usual rayon processes have been modified to meet specific uses. Strong viscose rayon is used for automobile tire cords because it is much stronger than cotton, especially at high temperatures. A viscose staple fiber that resembles wool has been developed. It has a rough surface and a permanent kink or crimp like wool. Animalized fibers are modified viscose fibers which possess wool dyeing properties due to the fact that some protein as casein has been added t o the spinning solution. Viscose to which synthetic resin has been added is similar to the animalized fibers, for it also has dyeing properties similar to wool. Teca is a cellulose acetate staple fiber which has a permanent kink or waviness and closely resembles wool. It can be used alone or in blends with other fibers to produce fabrics of wool-like appearance.

A new viscose fiber with permanent crimp, known as Fiber D, offers promise to rug manufacturers. It is round in cross section like wool, but has a smooth surface. It feels and has an appearance like that of coarse wool used in rugs and carpets.

FIBER IDENTIFICATION

It has become difficult t o ascertain the fiber content of fabrics by their appearance and handle or by chemical and burning tests. Formerly, it was a simple matter to determine whether a fabric was silk, cotton, linen, or rayon. Many of the new finishes give the synthetic fabrics the appearance of wool, linen, or silk. Furthermore, there has been an increasingly large production of fabrics composed of two or more kinds of fibers. Experienced textile workers have frequently been unable t o determine fiber content without using exacting methods of identification. Because the kinds of fibers used cannot be determined without reliable laboratory techniques, fabrics on the market should have labels stating accurately the fiber content. Such accurate statements of the fiber content would enable the consumer to exercise proper care in dry cleaning and laundering and would be of particular value to those who are allergic to certain textiles.

The textile technologist employs many tests to identify fibers. The burning tests are the easiest to use, but are not exacting, especially for mixed fabrics. Chemical tests are often helpful, but these are usually not applicable except in a chemical laboratory.

The microscopical identification of fibers is the most reliable and satisfactory. For this a microscope of high magnification must be available. Also, one must be equipped with the necessary apparatus to make mounts of the fibers both in longitudinal and crosssectional views. Viscose and cuprammonium rayons can be distinguished from cellulose acetate by the burning test. The acetate rayon melts to a crisp black bead, and the other rayons burn quickly like paper or cotton with no odour and no ash. Viscose and cuprammonium rayons are difficult to distinguish by chemical means, for they are both regenerated cellulose.

They are best identified by the shape of their cross sections. All three fibers appear like smooth glass rods, but the cross sections have characteristic shapes. The cuprammonium fiber is round; the acetate has a clover-leaf appearance with three or four lobes; and the viscose has sawtooth edges. Acetate rayon is easily distinguished from the natural and synthetic fibers because it is readily dissolved in acetone and the others, except vinyon, are not. Acetone is the base of nail polish remover and a drop of remover on an acetate rayon or vinyon fiber will dissolve it, but has no effect on the other fibers. Strong acids will attack acetate rayon, but will not attack vinyon.

The casein fibers cannot be distinguished from wool fibers by chemical or burning tests. They have practically the same chemical composition as wool and both have the odour of feathers when burned, leaving a black crisp ash. Both wool and casein fibers are easily dissolved in alkalies and are resistant to acids. The wool fiber is readily distinguished from other fibers under a microscope because it has scales on the surface which is not characteristic of any other fiber. The surface of lanital is smooth and the cross section is round. Soybean fibers react similarly to casein fibers and wool by chemical and burning tests. The fiber is smooth in appearance and has no scales.

The cross section is round. Both nylon and vinyon melt a t high temperatures and are practically non-flammable. They are easily distinguished by the shape of their cross sections. Nylon is round, and vinyon is shaped like a dumbbell. Vinyon is not attacked by either strong acids or alkalies; nylon is resistant to alkalies, but is injured by strong acids. Glass fibers are readily identified because when held in a flame they will not burn or melt, except at extremely high temperatures of over 1,000° F. All other fibers when held in a flame will either burn or melt. Glass fiber has a round cross section and is the smallest of all the textile fibers.

SPUN RAYON

The extensive use of rayon staple fiber has come about since 1930. Staple fiber is made by spinning rayon filaments and then cutting them after they leave the spinnerette into short uniform lengths, usually from about one t o six inches. These short lengths spun into yarn is known as spun rayon. Most of the staple rayon is made by the viscose process; it is also produced by the acetate and cuprammonium methods.

Textile mills spin staple rayon into spun rayon yarns by the cotton, wool, spun-silk or linen spinning systems. Fibers from one t o two inches long can be spun on cotton machinery; fibers cut into three-to-five inch lengths on the spun-silk system of spinning. If the rayon staple fibers are to be mixed with a natural fiber such as wool, they are cut into lengths approaching the length of the natural fibers.

Staple rayon really constitutes another textile raw material, and is a recent development of major importance in the textile industry. Spun rayon has increased in importance until fabrics made of this yarn are becoming more numerous and more in the fashion spotlight. Spun rayon has made possible a wide variety of fabrics.

It may be made and blended with wool, hair, silk, linen and cotton to form an infinite range of new and unique fabrics. Staple rayon may be combined with the natural fibers as well as with continuous filament rayon, When used alone or mixed with other fibers it can be woven into soft challies, heavy suitings, crepe fabrics, plushes and coatings.

These fabrics can be given both crisp and soft finishes. It can be made into fabrics resembling linens, cottons or woolens, in addition to an endless variety of new textures. They are made into fabrics such as hopsacking, light-weight sheers, smooth serges and poplins, gabardines, worsted-like suitings, shantung and poplin shirtings. Spun rayon may be of increasing importance in the manufacture of upholstery, draperies, decorative fabrics and domestics such as sheets, towels and tablecloths.

MIXED FABRICS

Rayon staple fiber has made possible the manufacture and increasing use of mixed fabrics. Combination of the natural fibers heretofore unmixed-with synthetic fibers has made available a greater variety of novelty fabrics and lower priced fabrics, since the new fibers are less expensive and more abundant than natural fibers. Two or more kinds of fibers can be mixed or blended before they are spun into yarn.

In this way a yarn is composed of different kinds of fibers blended before spinning. Yarns may be made of two or more single or ply yarns. One ply may be made of one kind of fiber and another of a different kind. Also, mixed fabrics may have the warp yarns of one kind of fiber and the filling of another. Tablecloths may be made of viscose rayon and cotton. Specimen A has a cotton warp and rayon filling; specimen B has two-ply yarns in both warp and filling.

The smaller ply shown raveled and untwisted is cotton and the larger rayon. Fabrics made of staple rayon fiber and wool have been made popular by manufacturers of women's suitings and cloakings, creating fabrics of increasing attractiveness which are lighter in weight than an all-wool material. The silk dress that the American woman has been wearing in past years has, in many cases, become a silk and rayon or an all rayon dress of today. The "silk trade" has now become a "fine fabric trade" which may completely disappear because of the inavailability of raw silk from Japan. For many years the silk flat crepe, sheer and satin were the product of the busy looms.

There was little need for showing originality in weave. The introduction of acetate rayon yarns, however, offered a wide field of new ideas, and silk manufacturers have recognized the possibilities of rayon in fine quality fabrics. A satin with a silk warp and cuprammonium rayon filling has replaced to a large extent, the all-silk satin which for years has been used for the manufacture of high-grade slips and pajamas. The other rayons have been used in combination with silk.

Cellulose acetate and viscose rayons are frequently combined. Many of the alpacas, crepes and satins which many people believe t o be silk are made of acetate and viscose rayons. In these fabrics the warp yarns may be acetate and the filling yarns viscose rayon. Yarns of two ply are often used, one ply of continuous filament acetate rayon with only a little twist and the other of a tightly twisted viscose rayon. Specimen D shows the fabric after the ply of acetate rayon has been removed. Some of the light weight dress and suit fabrics for women are mixed fabrics.

Hopsackings are made of viscose rayon and cotton and some of viscose rayon, silk and wool. Attractive dress fabrics are woven of ply yarns, one ply linen and the other cotton or one ply cotton and the other rayon. The smaller ply shown raveled and untwisted in both warp and filling is cotton and the larger viscose rayon. Rayon staple fiber is othen blended with linen. The light weight velvets usually have a silk back and a rayon pile. Suitable for dresses, has a viscose rayon pile and a silk back which can be seen through the pile. Woolen novelty fabrics are often made with wool blended with rabbit fur or feathers. Challies, alpacas, and flannels are made of blends of viscose rayon and wool. Many tweeds and novelty mixtures are made of wool and silk.

FINISHES USED ON SYNTHETIC FABRICS

Rayons in the 1920's had many shortcomings. At that time this shiny, sleazy fiber was regarded as a substitute for silk, and it was not desirable for many uses. Rayon garments would often fall to pieces when laundered. Dry cleaning and laundering caused the fabrics to shrink or stretch excessively. Yarn slippage and pulling a t seams were common complaints, and the high gloss of the rayon yarns was particularly objectionable for certain fabrics. Rayon fabrics have been made more useful by being given various finishes and treatments.

DELUSTERING

A dull finish to shiny rayon can be given by cutting it into staple lengths, by incorporating chemicals in the spinning solution or by coating the surface of the yarns after they are spun. In the first method when short lengths are spun into yarn the reflection of the light is broken up. Fabrics madc of continuous filament rayon is highly lustrous, and those of spun rayon are dull. Titanium dioxide, a white inert chemical substance added to the spinning solution, is used extensively for producing dull rayon.

CREASE RESISTANCE

Rayon fabrics, especially those made of spun rayons, crush and wrinkle easily just as cottons and linens. When these fabrics are allowed to hang, wrinkles and folds will not come out as they do in silk and wool fabrics which are resilient and thus have much resistance to creasing. Chemical treatment's to rayon fabrics have produced crease resistant finishes. These finishes impart a wooly handle, and as a result spun rayon can be changed into a wool-like material.

Synthetic resins are employed in treatments for crease resistance. The resins most popular a t present are the urea formaldehyde resins, a substance produced by dissolving urea, a white crystalline organic substance, and formaldehyde in water.

These chemicals are colourless and soluble in water, but when applied to the fabric with a heat treatment they coat the fibers with a transparent, thin film which is little affected by dry cleaning and laundering. Synthetic resins give many finishes to fabrics, depending upon how they are applied. They set dyestuffs and render fabrics more resistant to light and laundering. They also can be used to give a crisp, permanent starchlike finish.

PREVENTION OF YARN SLIPPAGE

The prevention of yarn slippage within the cloth has been brought about by the synthetic resins. At first rayons were often undesirable because of the great amount of pulling at the seams. Excessive fraying is eliminated as well by treating the fabrics with synthetic resins.

WATER-REPELLENT FINISHES

The substances used in the finishes to render fabrics resistant to water coat the fibers or cloth so they will not absorb moisture. If the fabrics are treated by the older methods in which wax, gums and metallic soaps are used they lose their water resistant property upon being laundered or dry cleaned because the finish is removed. These older finishes would not only make the fabrics treated impervious to water but to air as well. The new water-repellent finishes, produced by treating the fabrics with synthetic resins, are permanent t o washing and dry cleaning. The fibers are coated with a colourless resin which leaves the cloth porous to air. Dresses, coats, hosiery, draperies treated with this type of finish resist spotting by mud, beverages, ink and common stains. Textiles thus treated retain their original freshness.

FIRE PROOFING

Some textile fabrics such as viscose and cuprammonium rayons burn readily when held in a flame. Chemicals can be applied to these to render them fire resistant, forming a gas which keeps the oxygen away from the fabric, and consequently the flame is smothered. Other chemicals used in these finishes melt and form a non-flammable coating over the fibers. Some of these chemicals are removed by laundering and dry cleaning, and others can be made permanent. Fireproof fabrics will char but not blaze. Curtains, drapery materials and carpets are often treated to combat the fire hazard.

MILDEW PROOFING

Fabrics used and stored in damp climates and those that must withstand weathering are likely to mildew, especially in warm weather. Awnings and shower curtains often mildew. Chemicals can be applied to fabrics which will prevent mildew. Many of the mildew treatments are not permanent to laundering and dry cleaning.

SHRINK PROOFING

Cottons and linens have been given treatments for some time that would render them free from shrinking. This is done by wetting the fabrics and allowing them to dry when held under a small tension which is not sufficient to stretch them. Fabrics so treated are designated as preshrunk or "sanforized." The excessive stretching and shrinking of rayons cannot be overcome by this treatment. An outstanding development in the finishing of textile fabrics, however, has recently been perfected to shrinkproof rayons, and rayon fabrics can be made as shrinkproof as sanforized cotton. This is done with a chemical treatment of a synthetic resin and a modified sanforizing operation. This chemical finish will not wash out.

Most of these new finishes-shrinkproofing, waterproofing, crease resistant finishes, slip-proofing, delustering, fireproofing, and crisp starchlike finishes-are obtained largely by chemical treatments. The progress in the synthetic organic chemical industry is largely responsible for the advances made in textile finishes. By means of the great variety of new finishing materials made available by the chemist, the textile finisher has been able to produce effects heretofore unobtainable.

CARE OF SYNTHETIC TEXTILES

Garments made of rayon and other synthetic textiles will dryclean and many can be laundered successfully. The yarn in these fabrics can be cleaned with soap and water, but often the construction of the cloth and the dyestuffs prevent these fabrics from being laundered. If they rapidly lose their colour when laundered they should be dry-cleaned only. Some dyes are fast to dry cleaning, some to laundering and others only to sunlight. Few dyes are equally fast t o all treatments.

Rough crepes, sheers, georgettes, transparent, velvets, rough novelty fabrics and other materials often shrink excessively when wet. These should be dry-cleaned and not laundered. It is a good practice to experiment with a small piece of the fabric before the entire material is cleaned. If a small piece is laundered by hand it can readily be seen whether the fabric fades, shrinks or stretches unduly. If the fabrics are not washable they should be sent to a competent dry cleaner.

Soiled, spotted and stained garments should be cleaned as soon as possible in order to avoid "setting" of the stain. Dry cleaners should be informed of the kind of fiber of which a cloth is made and the type of stain in order to know what cleaning methods to use. Garments made of rayon or other synthetic fabrics which are washable should be given the same kind of care as those made of other fine fabrics.

Some synthetics have superior qualities, but many will not withstand abuse any better than the natural fibers. Many ready-to-wear garments carry a tag or label giving specific washing instructions. If no information is given the following suggestions may be followed for laundering most synthetic fabrics. A neutral soap and lukewarm, not hot, water should be used Dissolve soap thoroughly in water before fabric is dipped in it. Coloured garments should be washed separately and never soaked. Squeeze the sudsy water through the garment, and do not twist, wring, rub or handle it roughly, because rayons and most synthetics lose strength when wet but recover it when dry. Rinse thoroughly in lukewarm water.

Squeeze out moisture, roll in a towel to take out excess moisture, then unroll immediately. Stretch garment to proper dimensions and allow it to dry quickly in the shade. It is better to dry many garments, particularly knitted ones, on a flat surface. Never hang on sharp wires or pegs, and do not allow them to be blown by the wind. Press when nearly dry on the wrong side with a warm, not hot, iron. Many white viscose and cuprammonium rayon garments can be laundered in much the same way as cotton garments. They are weaker when wet, and care must be taken not to wring, pull, or rub them vigorously. They can be ironed a t as high a temperature as cotton. Acetate rayon or mixed fabrics containing acetate rayon must never be ironed with a hot iron. Rayon fabrics can be dry-cleaned in the home. The same precautions should be followed here as in any dry cleaning.

The ingredients of dry cleaning solvents used on acetate rayon should be known, because cleaning solutions containing chloroform, ether or acetone will injure the fabric. Special precautions should be followed in caring for some of the newer synthetic fabrics. Nylon hose can be washed and cared for just as silk hose. Lanital and other casein fabrics may be laundered in the same manner as wool. They lose strength when wet and must be handled gently. The fabrics should not be kept damp for a long time because they mildew quickly. Vinyon fabrics can be washed with soapy water not hotter than 149° F. At higher temperatures they will shrink and melt. Consequently, they must be pressed with a cool iron.

Fabrics of glass fiber can be easily laundered and quickly dried because they do not absorb moisture. These fabrics can be soaked and dipped up and down in hot soapy water. They should never be rubbed or wrung. The fabrics are rinsed by dipping them up and down in clear water to which has been added one or two tablespoons of olive or mineral oil. They should be smoothed out to dry on a flat surface. Wrinkles may be pressed out with a moderately hot iron. Glass fabrics should not be folded under pressure when stored, for the brittle fibers may break.

SERVICEABILITY OF SYNTHETIC FABRICS

The wide use of synthetic textiles has made it desirable for the consumer to know the relative serviceability of fabrics made of the synthetic and the natural fibers. Fabrics of the natural fibers are not always superior to synthetic fabrics. Tests can be made in the laboratory that will predict the usefulness of fabrics. Several studies have been made in the Department of Clothing and

Textiles at Kansas State College to ascertain the serviceability of synthetic fabrics.

EFFECT OF LIGHT AND HEAT ON THE COLOUR AND DETERIORATION OF RAYON FABRICS

It has long been known that light and heat will change the colour and decrease the breaking strength of textile fabrics. White silk and wool will become yellow in colour when kept in the light for long periods of time, and both can be scorched easily with a hot iron. Sunlight weakens silk considerably. It has been found that light and heat affect the colour and breaking strength of viscose, acetate and cuprammonium rayons Light does not decrease the strength of these fabrics as it does the strength of silk, however.

Exposures of 100 hours decreased the strength of the rayon only about one-tenth. Light had no effect on the colour of the white rayons, for they were equally as white after they were exposed to light. All three rayons turned yellow when placed in an oven a t 270-280° F., but none of them turned as much as silk or wool.

The acetate rayons changed less in colour and decreased less in breaking strength when exposed to heat than the viscose and cuprammonium rayon fabrics. Consequently, in cases where fabrics must be subjected to light and heat for many hours, rayons would be more serviceable than silk.

SERVICEABILITY OF SILK AND RAYON FABRICS

Although rayons have not been as serviceable as pure dye silks, they are generally considered as serviceable as is desired for dress fabrics in this period of rapidly changing fashions. Moderately weighted silks are also generally considered to have sufficient serviceability for satisfactory use as dress fabrics The serviceability of taffeta, flat crepe, and satin crepe fabrics made of pure dye silk, weighted silk; viscose rayon, and cellulose acetate rayon[2] was determined by laboratory tests. It was found that the breaking strength of viscose rayon was the greatest of the taffetas; the acetate rayon in the flat crepes; and pure dye silk in the satin crepes. The satin crepes shrank more than the taffetas and flat crepes.

Dry cleaning did not affect the breaking strength of any of the fabrics. Fabrics of no particular kind of fiber stood out as superior throughout the various tests. The perfection of finer rayon yarns has resulted in fabrics of extraordinary beauty and quality. Rayon yarns, except acetate, can be highly twisted and made into crepe, giving a deeper pebble and a greater variety of crepe effects than is possible with silk. The rayon crepes are difficult to distinguish from silk crepes in appearance and handle. The comparative serviceability of viscose rayon crepes with silk crepes was studied. The breaking strength of the silk crepes was approximately twice as large as the rayon. Also, the silk fabrics had greater elongation, greater resistance to creasing and less shrinkage than the rayon. Dry cleaning did not affect the breaking strength or the colour of either. Both possessed a comparable degree of colour fastness to light, but the silk crepes showed greater resistance to fading from perspiration. None of the fabrics showed any slippage.

There was no relationship between the price of the material and the durability of the fabrics. This study showed that the pure dye silk crepes were more serviceable than the rayon. Rayon is being used in the manufacture of all rayon satin and crepe suitable for women's slips. Crepe and satin fabrics are also being made with silk warp and rayon fillinga. Crepe and satin slip fabrics of all silk, all rayon and rayon and silk mixed have been tested.

The breaking strength of the silk fabrics was greater than the rayon and also greater than the mixtures. Laundering did not decrease the breaking strength of the slip fabrics. The rayon fabrics shrank more than the silk or mixtures. The appearance and handle of the fabrics showed a change after ten launderings and even greater change after twenty launderings. The all silk satin showed only a slight loss of colour, but the all or part rayon satins and crepes faded to a marked degree.

The white silk showed a tendency to turn yellow after launderings, whereas the all rayon crepes retained their original whiteness. The all silk satin and crepe lost much of their body after the laundering, probably because of the loss of soluble weighting. The all- or part-rayon fabrics lost some of their softness after the launderings. The all rayon fabrics acquired harsh handle that lessened their desirability. The all rayon satin was the only fabric which lost its silk-like appearance. This satin became shiny, and lost entirely its softness after the laundering. The silk-and-rayon mixed fabrics retained their body and pleasing handle. The only unsatisfactory change in these mixed fabrics was the loss of colour.

SERVICEABILITY OF WOOL, RAYON, AND MIXED WOOL-AND-RAYON FABRICS

Significant, differences have been found in the characteristics of wool, rayon and mixed wool-and-rayon fabrics. The materials studied were dress fabrics of all wool, all rayon and wool and viscose rayon mixtures. The all rayon and mixed fabrics had greater breaking strength than the wool, and the mixtures were not affected as much by abrasion as either the wool or the rayon. Dry cleaning did not affect the breaking strength.

The wool had the greatest elongation and rayon the least. The difference in shrinkage of the wool and rayon was not significant. The mixed fabrics shrank much more than either the wool or rayon. The wool and mixed fabrics were similar in resistance to creasing, and both were much more crease resistant than the rayon. The appearance of all the fabrics was satisfactory after dry cleaning. None showed change in colour except the white wool which became somewhat yellow.

SERVICEABILITY OF COTTON, LINEN, AND RAYON MIXED FABRICS

Spun rayon is used for the production of fabrics similar in appearance to linen, and it is also mixed with cotton and linen for dress materials which can be laundered. In a study of all linen, all viscose rayon, and viscose rayon and linen mixed fabrics it was found that the breaking strength of the linen was greater than the rayon or the mixtures. There was enough linen in the mixed fabrics to diminish the decrease in the wet breaking strength. Thus, these fabrics of linen and rayon had a greater wet breaking strength than those of all rayon. The linen was least resistant to wear. The mixtures shrank the most and the linens the least.

The average shrinkage in the warp of the mixtures and rayon fabrics was 8.6 and 6.0 percent respectively which was excessive; and that of the linen was only 2.7 percent. All the fabrics retained their colour and had a desirable appearance after laundering. The laundering decreased the breaking strength. All viscose rayon and viscose rayon-and-cotton mixed fabrics have been compared in breaking strength; the rayon was stronger than the rayon-and-cotton mixed fabrics, and both groups were equally resistant to wear. There was enough cotton in the mixed fabrics to diminish the decrease in the wet breaking strength. The average shrinkage in the warp of rayon and rayon and cotton mixtures was 4.1 and 3.5 percent, respectively. Both groups of fabrics laundered satisfactorily. Laundering did not diminish the breaking strength of these fabrics.

THE EFFECT OF RESIN FINISHES ON RAYON FABRICS

Two groups of viscose rayon gabardines were tested for their One group had a synthetic resin finish, and the other had not. The resin finishes were not removed by dry cleaning or laundering. The resin finished fabrics faded slightly, but many of the non-resin finished fabrics showed marked degrees of fading. In laundering the shrinkage of the gabardines without resin finish varied from 6.0 t o 13.0 percent, and those with resin finish varied from 0.3 to 4.0 percent. The two groups showed differences in breaking strength. The ones with the resin finish had much greater breaking strength.

The treated fabrics had a firmer handle than the untreated fabrics and had a better appearance after dry cleaning and laundering. These fabrics did not become sleazy after repeated dry cleanings and launderings. The serviceability of the rayon gabardines was considerably increased by resin finishes.

CONCLUSION

Synthetic fibers were first made to simulate silk. Textile technologists endeavored to produce fibers with characteristics of the natural fibers, silk, wool, cotton and linen. Now they are attempting not to imitate these fibers, but to create fibers with characteristics which are distinctive from the natural fibers. The natural fibers have definite physical and chemical properties which cannot be changed a t will, but the synthetics can be made with a variety of properties. Some of the newer synthetic fibers have properties quite different from the natural fibers. No doubt others will be created having still more varied characteristics. Improvement of production has enabled the textile manufacturer to vary the properties of synthetic fibers. The fibers and yarns are uniform when produced under exacting conditions. The size of filaments has been varied and made considerably smaller.

The finer yarns of various lusters have aided in the almost universal use of rayons for fine dress goods. Many designers and dress manufacturers have chosen a particular construction in rayon rather than silk. Versatile rayon has made possible the development of many new fabrics that never existed before. This has been an inspiration to the designer. The synthetic fibers not only make possible fabrics of great variety, but some also have qualities that are superior to the natural fibers. Many of the synthetic textiles have been in use long enough to prove their desirable qualities which have made them preferred to those of the natural fibers. If a textile fiber with certain properties is needed, the scientist will endeavor to create it ; and judging by past achievements he will be reasonably successful. No one dares to predict the appearance, characteristics and uses of the new synthetic textiles of the future, or from what raw materials they will be made.

5

Knitting

TECHNIQUES

BASKETWEAVE (KNITTING)

In knitting, a basketweave pattern is characterized by intersecting ribs and welts. In its simplest form, basketweave is composed of a checkerboard pattern of identical rectangles that alternate between stockinette stitch and reverse stockinette stitch. In this form of basketweave, the only variables are the dimensions of the rectangles, *i.e.*, the number of stitches across and number of rows long. If the rectangles are much longer vertically than horizontally, a basketweave may also be called a broken rib pattern. Similarly, if the rectangles are much longer horizontally than vertically, a basketweave may also be called a broken welt pattern. More complicated types of basketweave can be made with allowing the rib and welt widths to vary (*i.e.*, by using rectangles of different sizes), and by allowing some rows or welts to continue uninterrupted.

Other variations of basketweave change the stitch patterns that compose the rectangles. Instead of stockinette stitch and its reverse, the rectangles can be done in seed stitch, in cables or lace. An especially common variant is to alternate rectangles of ribbing with rectangles of welting, *e.g.*, rectangles of three ribs set against rectangles of three welts. An visually fascinating alternative approach to basketweave patterns is entrelac knitting, in which the grain of the knitted fabric itself follows a woven pattern.

BEAD KNITTING

Beaded knitting is a type of knitting in which the stitches are decorated with ceramic or glass beads. The techniques are described here for beads, but knitting sequins (and other perforated objects) can be done analogously.

Techniques

There are two main approaches- threading the beads on in advance, and threading the beads on individually as the object is knitted.

Threaded on in advance

The classic approach is to thread beads on the yarn in advance. There are several ways to go about knitting the beads in though: slipping the stitch, putting the bead between stitches, and putting the bead on the stitch.

The slip stitch method is to slip the stitch with the yarn (and bead) in front at the position where a bead is desired.

The bead will be visible only on the right side of the work. The disadvantage of this method is that beads cannot be arranged on top of each other, since the same stitch would have to be slipped multiple times. The axis of the bead is horizontal to the work.

Putting the bead between the two stitches is done by positioning the bead on the yarn connecting two stitches, *i.e.*, between two bights. This results in the bead being visible from both sides of the work when beads are slipped on both sides. The bead has a tendency to not lie straight when placed with this method. The axis of the bead is horizontal with respect to the work. This is the method shown in the picture.

The third method is to knit the stitch with the bead on the bight itself. Technically, this is considered "bead knitting" rather than "beaded knitting." For consistency, the bead should be positioned on the same leg of the bight. Using this third method, one can make a densely beaded knitted fabric, *i.e.*, one that appears to be all beads, with no knitted yarns visible. It is sometimes difficult, however, to keep the bead on the right side of the fabric; for this reason, the stitches are often twisted, to tighten up the fabric.

Threaded as it is knitted

A second approach that offers more flexibility is to thread the bead onto the bight, *i.e.*, onto the loop before it is knitted. (The rotational axis of the bead is aligned vertically.) In this case, the beads appears on *both* sides of the fabric. The chief drawback of this method is that the beads must be larger, since two strands of the yarn pass through the bead. In this approach, it is helpful to make an elongated stitch to fit both the bead and the next stitch.

A third approach is to sew on the beads after knitting. This approach is the least desirable, since it does not protect the yarn as well and since the sewing thread is usually weaker than the yarn.

History of Bead Knitting

In the Victorian era, it was popular to knit densely beaded purses using tiny knitting needles made of stiff wire, much smaller (less than 1 mm) than the smallest knitting needle easily available commercially today. Such tiny stitches allow for very intricate patterns. Very small lace needles, as small as 0.5 mm diameter, are available from specialty retailers.

BIAS KNITTING

Bias knitting is a method by which the grain of a knitted fabric is skewed from the vertical, by decreasing on one side and increasing on the other. The term "bias" derives from sewing, where bias refers to the diagonal between the weft and warp threads, where the elasticity of wovens is greatest.

A classic use of bias knitting is in making a saw-toothed knitted hemline for a garment, with a central double decrease flanked by single increases on either side.

BINDING OFF

In knitting, binding off, or casting off, is a family of techniques for ending a column (a wale) of stitches. Binding off is typically used to define the final (usually upper, taking the cast on edge as the lower) edge of a knitted fabric, although it may also be used in other contexts, *e.g.*, in making button holes. In principle, binding off is the opposite of casting on, but the techniques are generally not mirror images of one another. Sometimes, however, they can produce a mirror image appearance.

Techniques

Binding/casting off generally involves passing the final loop of a wale over the loop of an adjacent stitch. Techniques differ, however, in how this is done:

- Simple bind/cast off : Involves simply passing each loop over an adjacent stitch. (The yarn is passed through the final loop to secure the whole chain.) This technique produces a tight edge with little elasticity. Its advantages are that
  - — it does not flare as many bound/cast-off edges do;
  - — it will retain its length, even under tension (good for shoulder seams);
  - — it does not require that the knitting yarn be nearby;
  - — it can be done in any direction.
- Knit bind/cast off: Involves knitting each loop before passing it over the next loop. Note that you start by knitting two stitches as normal, then pull the first knitted stitch on the right needle over the second knitted stitch on the right needle, knit the next stitch on the left needle, and continue to the end. This can be done tightly or loosely, depending on the tension of the knitter. If you want to ensure that the bind/cast-off edge is loose, one method to ensure this is to use a needle at least two sizes larger than the project needle size.
- Purl bind/cast off : Involves purling each loop before passing it over the next loop. This is simply the purl version of the knit bind/cast off. You can also do this in pattern, such as a rib pattern.

- Decrease bind/cast off : Involves iterative decreases, *e.g.*, "k2tog, return loop to left needle".
- Sewn bind/cast off : This involves the use of a tapestry needle with an extremely long tail of the yarn. There are several variations. The most well-known is from Elizabeth Zimmermann's *Knitting without Tears*. It produces an extremely stretchy edge and is excellent for the top of socks. Using this long tail, pull the needle through two front loops as if to purl, then back through the first stitch as if to knit. Drop the first stitch off. Repeat to the end.
- Three needle bind/cast off : This bind/cast off is used for joining two pieces of knitting, such as the front and back of a sweater by binding/casting off their stitches together to create a seam. For this bind/cast off, the two needles (in the case of a sweater, each may hold the shoulder stitches) are held parallel with the right sides of the knitting facing each other. Binding/casting off as with the knit bind/cast off, each stitch is the result of knitting together one stitch from each needle (*i.e.* pass needle through first stitch of first needle, first stitch of second needle, wrap yarn, pull through, and drop both stitches)
- Tubular bind/cast off : A stretchy bind/cast off used on a ribbed edge; most helpful for sock and neckline edges.

BOBBLE (KNITTING)

In knitting, a bobble is a localized set of stitches forming a raised bump. The bumps are usually arranged in a regular geometrical pattern (*e.g.*, a hexagonal grid) or may be figurative, *e.g.*, represent apples on a knitted tree. The basic idea of a bobble is to increase into a single stitch, knit a few short rows, then decrease back to a single stitch. However, this leaves many choices: how to increase and how many stitches, how many short rows to work, and how to decrease. A bobble can also be a yarn Pom-pon used to decorate knitted items such as Bobble hats.

BRIOCHE KNITTING

Brioche knitting is a family of knitting patterns involving tucked stitches, *i.e.*, yarn overs that are knitted together with a slipped stitch from the previous row. Such stitches may also be made by knitting into the row below (equivalent to the slipped stitch) and dropping the stitch above (equivalent to the yarn over).

The tucked stitches may form a second layer of knitting in front of the first layer, resembling an array of arches or (seen upside-down) of fish scales.

Although warm and beautiful in texture, garments made from brioche knitting have the possible drawback that the tucked stitches might get caught and be pulled out.

Brioche knitting may have originated in the Middle East. However, the term "brioche" seems to have derived from French slang for "mistake". The name might be a reference to the brioche dinner roll, which is formed of two pieces, one stacked atop the other.

Brioche Stitch is included in Barbara G. Walker's *A Treasury of Knitting Patterns* and in Elizabeth Zimmermann's *Knitting Without Tears.* Nancy Marchant introduced several varieties of the brioche stitch to American knitters with her book *Brioche Knitting,* inspired by the brioche knitting she found very prevalent in the Netherlands. The brioche stitch can be used to knit any kind of garment or project that regular knitting can be used for, but will be double-thick. Nancy Marchant standardized brioche-knitting abbreviations and terminology so that knitters world-wide could share patterns and understand the abbreviations.

Marchant has attempted to standardize and simplify the terminology for the stitch. She defines the basic stitch of brioche knitting as the brioche-knit stitch, which she calls the "bark" stitch (abbreviated "brk"), which consists of a knit-stitch knitted together with its "wrap," a yarnover from the previous row. The brioche-purl stitch (or the "burp" stitch (abbreviated "brp") is the purled version.

Each bark or burp stitch is followed by a yarn-front, slip-one, yarnover (yf-sl1yo). This sets up the bark and burp stitches for the next row.

In brioche knitting, it takes two "passes" to complete a single row of knitting, since only half the stitches are knitted each time. The other half are slipped. For this reason, it takes more knowledge to be able to count rows and stitches and measure gauge.

BUTTONHOLE (KNITTING)

In knitting, buttonholes can be made in several ways.

Buttonhole Techniques (knitting):

- Eyelet buttonhole — Work to desired buttonhole location, yarn over, decrease.
- Vertical buttonhole — Knit to desired buttonhole location, attach a second yarn and begin knitting with this yarn; knit the two fabrics flanking the buttonhole independently until the work reaches the desired length of the buttonhole. Rejoin the two flanks, drop the other yarn end, and work as one again.
- Two-row buttonhole (horizontal) — Bind off an appropriate number of stitches and cast them back on in the next row. The yarn is usually woven into the bound off stitches, for reinforcement.
- One-row buttonhole (horizontal) — Work to where you want the buttonhole to be. Slip next stitch, move yarn to the back of work,

slip second stitch, pass first slipped stitch over the second, slip next stitch, pass second stitch over the third, and repeat until desired width of buttonhole is reached. Turn work and, by using the cable cast-on or backward loop cast on, cast on the number of stitches bound off plus one. Turn work again. Slip the next stitch to the right-hand needle and pass the extra cast-on stitch over the slipped stitch.

Any type of knit buttonhole can be strengthened (and given neater edges) with embroidery, specifically the buttonhole stitch.

CASTING ON (KNITTING)

In knitting, casting on is a family of techniques for adding new stitches that do not depend on earlier stitches, *i.e.*, stitches having an independent lower edge. In principle, casting on is the opposite of binding off, but the techniques involved are generally unrelated.

Casting on can also be decorated with various stitch patterns, especially picots. The cast-on stitches can also be twisted clockwise or counterclockwise as they are added to the needle; this is commonly done for the *single cast-on* described above to give it a neater, more uniform look.

Casting on is sometimes done with two needles, or a needle of larger size; the extra length of yarn in each stitch gives the edge more flexibility.

When casting on at the beginning, one end of the yarn must be secured to the knitting needle by knotting it, usually with a slip knot. This knot is unnecessary when casting on in the middle of the fabric (*e.g.*, when making the upper edge of a buttonhole) since the yarn is already secured to the fabric. The original slip knot can also be pulled out (after a few rows have been knitted) without damaging the knitted fabric. Once one loop has been secured around the needle, others can be added by several methods.

Methods for Casting on in Handknitting

- Knit-on cast-on : Perhaps the most straightforward method, in which a new loop is drawn through the previous loop and then added to the needle. However, this method is deprecated for giving an untidy edge. It can also be done in a purl version, or even a rib version.
- Cable cast-on : A closely related technique, in which a new loop is drawn through the space between the two previous loops and then added to the needle. This edge is firm and has a neat, corded look; although it may be too bulky with thick yarns.
- *Single cast-on : An even simpler method, also called the* simple cast-on *or "backward loop cast-on," which involves adding a series of half hitches to the needle. This creates a very stretchy, flexible edge. It is a common approach for adding several stitches to the edge in the middle of a knitted*

fabric, but it is difficult to knit from and make even. A variation is the twisted simple cast on, where you twist the new loop around your thumb, with the yarn going around the back of your thumb to the front as in the simple cast-on, but picking up the new loop from the back side of the loop. This is tighter and neater, but has less elasticity.

- Double cast-on : A common method, in which all the loops are made with one yarn, while the other end (the dangling end from the original slip knot) is used to secure the base of each loop. This method is also called the "knit half-hitch cast on" or the "long-tail cast on." Although popular, this method requires that the knitter estimate the length of the dangling yarn before the stitches are cast on; if the dangling yarn is too short, the knitter will run out of yarn with which to secure the stitches before the full number of stitches have been cast on. In that case, the knitter will have to pull everything out, re-position the slip knot to give a longer tail, and begin anew. Despite this shortcoming, it's a good all-around method for casting on. Another variation for this method is to use two different yarns, one the main yarns that you are using for your project, and the second a piece of contrasting waste yarn. You attach the two with a slip knot, and then using the waste or contrast yarn as your long tail, start your row. This is useful if you need to pick up stitches on your cast on edge in order to knit in the opposite direction. You can also use it decoratively, making the contrast or waste yarn a part of your pattern design. To execute it, you start by figuring out how much yarn you expect your cast-on row to require and pulling out that amount of yarn. Once you have that, put a slip knot on the needle (this is not absolutely necessary, since the first cast-on stitch will create a slip knot for you in the process, but it is generally more secure to start with a slip knot). Hold the needle in your right hand and the yarn in your left, with the long tail pulled around your thumb and hanging in front, and the yarn from ball around your first or second finger, with the ball tail heading towards the back. Once you have that, take the needle under the front of the long tail, picking up a half hitch, then back to the yarn over your finger from the top side of the yarn, pulling the loop through the half hitch you had formed.
  This cast on can also be done in a purl and a twisted stitch version as well.
- Tubular cast-on : Involves knitting onto a cast on row knitted in a contrasting yarn with half as many stitches. Each knit stitch into the contrasting stitches is followed by a yarn-over to double the number of stitches. After several rows, a tuck is formed by knitting together

the first and third rows, forming a tube through which elastic can be pulled. A neat edge, nicely suited for 1x1 ribbing.

- Provisional cast-on (standard) : Also known as an "invisible cast-on," since the waste yarn used can be pulled out later to allow the knitter to continue the knitting in the opposite direction. This cast-on is also the best method for double-knit fabrics, since the knitting has no boundary; the knitting is continuous from one side of the fabric to the other. Holding the ends of a waste yarn and the working yarn, make an overhand knot. Place a needle held in the left hand between the two yarns, with the knot below, the waste yarn held underneath and parallel to the needle out to the right, and the working yarn up and in front of the needle. Bring the working yarn down behind the needle and in front of the waste yarn; up behind the waste yarn and over-and-up then down in front of the needle; down behind the waste yarn; then up in front of the needle. Repeat for each two stitches. When desired number of stitches is reached, loosely fasten the waste yarn and work as usual with the working yarn. To take out the provisional cast-on, unfasten the end of the waste yarn and carefully pull it out, picking up the now loose loops on a needle and working from the opposite direction of previous work.
- Double needle cast-on : First, you must put all of your stitches on one needle.
- Braided cast-on : Frequently used in mitten edges ...
- Chain cast-on : Uses a crochet hook or two knitting needles. To execute, hold a knitting needle in left hand and crochet hook or second knitting needle in right hand. Make a slip-knot in yarn and put it on the crochet hook or right-hand needle. Wrap the yarn from the back of the left-hand needle and over to the front, over the crochet hook or right needle, pass the slip-knot loop over the wrap, leaving the new loop on the crochet hook or right needle. Repeat, wrapping the yarn over the left-hand needle before passing it over the crochet hook or right needle to make a new loop, until you have one less stitch than required. Place the last loop on the left-hand needle as the first stitch that will be worked. This cast-on creates an edge that mimics a standard bind-off edge.
- Crochet chain cast-on : For this you need to know how to do a simple crochet chain. Once you have chained enough to equal the number of stitches you need, plus a few extra, turn the chain over so that you see the bumps that were forming as you pulled the yarn through the hole. Put your knitting needle through those bumps and knit through it as normal. This produces the same edge as knitting on.

- Provisional chain cast-on : Simply the crochet chain cast-on using waste yarn; this is also an "invisible cast-on" that can be pulled out later to allow knitting in the opposite direction. Work a crochet chain in waste yarn, loosely fastening the tail end. With working yarn, pick up the chain-bumps, as for the crochet chain cast-on, to create the working stitches. To take out the cast-on, simply pull out the tail of the waste yarn at the fastened end and "zip off" the crochet chain. Pick up the now loose loops and work from the opposite direction of previous work. This is done in toe-up socks and shawls or scarves with directional patterns that need to start from a center edge.
- Turkish cast-on : Used for circular beginnings, often for the toes of socks made toe-up. It is invisible (as with the provisional cast-on). You begin with two circular needles held one above the other as you look at them (Upper called A, lower called B). Place a slip-knot on B, and wrap the yarn up behind A. Then begin wrapping it around both needles, down in front and up in back, until the number of wraps equals half the number of stitches needed. Slide B along, through the wraps, until they sit on the cable, and the ends dangle on either side. Then bring the other, loose, end of A up, and knit into the wraps still on A. Once you've knit all of those wraps, pull A until the wraps are on the cable, and pull B so that the tip of the needle holds the wraps, pointed to the end with the working yarn. Bring up the other end of B and knit across the wraps again. This completes one round. From here, you can continue to work around the stitches on the two circular needles, increasing as desired, or switch to double pointed needles or a single circular needle for the Magic Loop method of knitting circularly.
- *Magic cast-on : Developed by Judy Becker; also known as the "magic toe-up cast-on," due to its popular use in beginning toe-up sock construction. Instructions were first published in an issue of the on-line knitting magazine* knitty.com.
- Circular cast-on : Popularized by Elizabeth Zimmermann as "Emily Ocker's Circular Beginning."
- *Old Norwegian cast-on: Also known as the "German Twisted cast-on" and similar to the "long-tail cast on" but uses a longer tail due to a second twist in the thumb loop. Leaving a tail the necessary length, make a slipknot and place it on a needle held in your right hand. The slipknot counts as the first stitch. Place the thumb and index finger of your left hand between the yarn ends so that the strand connected to the ball is around your index finger and the tail end is around your thumb. Secure the yarn ends with your other fingers and hold your palm upwards,*

making a V of yarn. Bring the needle in front of your thumb, under both yarns around the thumb, down into the center of the thumb loop, back forward, and over the top of the yarn around your index finger. Use the needle to catch this yarn, then bring the needle back down through the thumb loop, turning your thumb slightly to make room for the needle to pass through. Drop the loop off your thumb and place your thumb back in the V configuration while tightening up the resulting stitch on the needle. Instructions published on knittingdaily.com.

DOUBLE KNITTING

Double knitting is a form of knitting in which two fabrics are knitted simultaneously on one pair of needles. The fabrics may be inseparable, as in *interlock knitted fabrics*, or they can simply be two unconnected fabrics. In principle, an arbitrary number n of fabrics can be knitted simultaneously on one pair of knitting needles with yarns, as long as one is careful. The most famous sample of double knitting is the pair of socks knitted simultaneously on one set of knitting needles by Anna Makarovna, the nanny in Leo Tolstoy's *War and Peace*

- When the pair was finished, she made a solemn ceremony of pulling one stocking out of the other in the presence of the children.

Double knit fabric became popular within 1970s fashion; in the 1989 Grammy Award–winning song "Parents Just Don't Understand", Will Smith comically excoriates his mother for forcing him to wear outdated, 1970s-era "double-knit reversible slacks". However, the double-knit fabric referred to is a machine-made double-thick construction with some similarities to handmade double knitting (durability, reversibility) but none of the unique colourwork and construction possibilities.

Methods

There are several methods for double knitting, including flat knitting on double-pointed knitting needles; after one row has been worked with one yarn, the knitter slides the stitches to the other end of the needle and begins the same row with the next yarn. Only half the stitches are knitted with any one yarn; the rest are slipped. After both passes are done, the knitter then turns the work and begins another row.

Another common method is to alternate a knit stitch of yarn A with a purl stitch of yarn B. Since the yarn is held to the back for a knit, and to the front for a purl, this results in two sheets of stockinette stitches, with the wrong (purl) sides facing each other. Switching colours ties the two sides together for a single double-thick fabric. This method is often used for elaborate two-colour designs, as there are few constraints on how the colours may be used. The finished item from this method is reversible, each side holding the negative image of the other.

DROP-STITCH KNITTING

Drop-stitch knitting is a knitting technique for producing open, vertical stripes in a garment. The basic idea is to knit a solid fabric, then (deliberately) drop one or more stitches (*i.e.*, draw a loop out from the loop below it, and so on repeatedly), producing a run (or ladder) in the fabric. The run will continue to the bottom (*i.e.*, cast-on) edge of the garment, or until it encounters an increase, at which it stops.

Drop-stich runs are an easy way to get the "see-through" effect of lace, but with a much more casual look. The designer can make interesting arrangements of open stripes (of whatever length); the endpoints of the stripe are specified by the initial increase and the point at which the first stitch was dropped. Drop-stitch stripes are usually vertical, since they follow the grain of the knitting, *i.e.*, the *wales*, the columns of dependent loops. However, the grain of the knitting can be made off-vertical, *e.g.*, in entrelac or by increasing on one edge and decreasing on the other. A drop-stitch that does not produce a vertical strip(or a run) occurs when the initial increase happens in the middle of a stitch (wrap the yarn over the right needle multiple times during the stitch which should produce what looks to be multiple yarn-overs on the right needle) and the increase is then dropped instead of worked in the next row. This produces a stitch that is twice as long as a normal stitch. This method will produce a horizontal stripe when it is used on multiple adjacent stitches. When the number of increases applied to a single stitch is raised the resulting stitch is proportionately lengthened.

Instead of being left open, the cross-strands of runs (the "rungs" of the ladder) can also be modified in various ways. For example, using a crochet hook, one can re-work them into larger knitted bights, *e.g.*, by drawing four strands through the four below them, and so on indefinitely. As another example, the "rungs" can be bound up in different patterns using a contrasting yarn and a darning needle.

WELTING (KNITTING)

In knitting, welting is the horizontal analog of ribbing; that is, one or more horizontal rows of knit stitches alternating with one or more rows of purl stitches. The simplest welting is garter stitch, in which knit rows alternate with purl rows. If the fabric is produced "in the round", the effect is simply produced by knitting one row, then purling the next. If the fabric is being knit back-and-forth, turned after every row, the effect is produced even more simply by knitting each row—first from the right side, then from the wrong side.

Similar to ribbing, a welting pattern can be specified by the number of knit rows followed by the number of purl rows, *e.g.*, 1x1 welting is garter stitch. Many complicated patterns of purely horizontal stripes are possible, which can be worked in yarns of different colour, thickness and texture for added visual interest, *e.g.*, wide stripes of red chenille alternating with narrow stripes of black worsted wool.

Welts can be decorated with nearly any motif used for a plain knitted fabric, *e.g.*, bobbles, lace, and various colours. However, cables cannot be made horizontally; if desired, cables must be worked as a separate piece. A tuck can be created when a previous row is knit together, stitch by stitch, with the present row, forming a round ridge that projects outwards towards the right side. Used as a decorative detail.

WEAVING (KNITTING)

In knitting, weaving is a family of techniques used for several purposes in knitting. The most common use for the technique is when working stranded colour patterns, in which two yarns are alternated to certain of the stitches to create patterns. Weaving is used to attach the horizontal strands of yarn that pass unused behind stitches formed with the other yarn to the inside of the fabric. This is usually done to prevent overly long "floats" on the inside to prevent snags and create an even tension in the fabric.

The technique can also be used for decorative purposes if it is done on every stitch, in which case the "inside" of the fabric is used on the outside

Weaving in, or "inlay", is a related but different technique that is used to thread an extra yarn(s) into the fabric without knitting it. The woven yarn(s) need not be the same thickness or colour as the knitted yarn, and almost always (but not necessarily) follow the horizontal rows (courses) of knitting. Because the extra yarn simply passes back and forth between the knitted stitches, it can be pulled out unless it is secured in some way at the sides. The resulting fabric is more like a woven texture than a knit because the extra strand reduces its elasticity.

Methods

To weave in for stranded colour patterns, the yarn being used to form stitches is passed first below and then above the other as it is wrapped around the needle, which catches the yarn not in use against the inside of the fabric.

For the Inlay method of weaving the extra yarn(s) is passed in back or in front of each stitch as it is knitted, similar to the slip-stitch knitting technique.

In another technique, dip stitches are used to secure the woven yarn(s). The method may be necessary if the woven yarn(s) are much thicker than any one row of knitting.

Other Meanings of "Weaving" in Knitting

Weaving has at least two other meanings in knitting. First, *weaving in the ends of the yarn* means burying the free ends of the yarn(s) in the knitted fabric to secure them, by passing them in and around the stitches. Second, *weaving* is sometimes used as a synonym for grafting.

GAUGE (KNITTING)

In knitting, the word *gauge* is used both in hand knitting and machine knitting; the latter, technical abbreviation GG, refers to "Knitting Machines" fineness size. In both cases, the term refers to the number of stitches per inch, not the size of the finished garment. In both cases, the gauge is measured by counting the number of stitches (in hand knitting) or the number of needles (on a knitting machine bed) over several inches then dividing by the number of inches in the width of the sample.

Gauge on knitting machines

There are two types of classification of Knitting Gauges or Unit of Measure:

- A – Used for Cotton Fully fashion flat machines (Bentley – Monk, Textima, Sheller etc..) where "Gauge" is measured in 1,5" Inches (2,54 cm x 1,5) and the machine's gauge is expressed by the number of needles needed to achieve that gauge.
- B – Used for hand, mechanical or modern Electronic Flat Machines (Stoll, Shima, Protti etc..), where gauge is measured in 1 inch increments (or 2,5 cm) and the machine's gauge is similarly measured by the number of needles required to achieve that number.

Compared graduation scale Gauge (GG) A versus B system: A 30 GG (A) Cotton Fully fashion flat machine (30 needles in 1,5") is comparable to a 20 GG (B) Electronic Flat machine, a 27 GG (A) is an 18 GG (B), an 18 GG (A) is a 12 GG (B), a 12 GG (A) is an 8 GG (B), a 7,5 GG (A) is a 5 GG (B) and a 4,5 GG (A) is a 3 GG (B).

Factors that affect knitting gauge

The gauge of a knitted fabric depends on the pattern of stitches in the fabric, the kind of yarn, the size of knitting needles, and the *tension* of the individual knitter (*i.e.*, how much yarn he or she allows between stitches).

- For example, ribbing and cable patterns tend to "pull in," giving more stitches over an identical width than stockinette, garter, or seed stitch. Even the same stitch produced in two different ways may produce a different gauge.
- Thicker yarns with less loft generally produce larger stitches than thinner yarns (reducing the number of stitches per width and length).

- Larger knitting needles also produce larger stitches, giving fewer stitches and rows per inch; changing needle size is the best way to control one's own gauge for a given pattern and yarn.
- Finally, the knitter's tension, or how tightly one knits, can affect the gauge significantly. The gauge can even vary within a single garment, typically with beginning knitters; as knitters become more familiar with a stitch pattern, they become more relaxed and make the stitch differently, producing a different gauge.

Sometimes the gauge is deliberately altered within a garment, usually by changing needle size; for example, smaller stitches are often made at the collar, sleeve cuffs, hemline ribbing or pocket edges.

Uneven knitting

Uneven knitting is a knitting technique in which two knitting needles of different sizes are used. The method is sometimes used when the knitter has a significantly different gauge on knit and purl stitches. It is also useful for producing elongated stitches and certain specialty patterns.

Knitting Gauge in Patterns

To produce a knitted garment of given dimensions, whether from one's own design or from a published pattern, the gauge should match as closely as possible; significant differences in gauge will lead to a deformed garment. Patterns for knitting projects almost always include a suggested gauge for the project.

For illustration, suppose that a sweater is designed to measure 40" around the bustline with a gauge of 5 st/inch in the chosen stitch. Therefore, the pattern should call for 200 stitches (5 st/inch x 40") at the bustline. If the knitter follows the pattern with a gauge of 4 st/inch, the sweater will measure 50" around the bustline (200 st / 4st/in) — too baggy!

Conversely, if the knitter follows the pattern with a gauge of 6 st/inch, the sweater will measure ~33" around the bustline (200 st / 6st/inch) — too tight! Generally, the gauge should match to better than 5 per cent, corresponding to 1" of ease in a 20" width. Similar concerns apply to the number of *rows per inch*.

Luckily, the gauge can be adjusted by changing needle size, without changing the pattern, stitch, yarn, or habits of the knitter. Larger needles produce a smaller gauge (fewer stitches per inch) and smaller needles produce a larger gauge (more stitches per inch). If necessary, further adjustments can be made by subtly altering the pattern dimensions, *e.g.*, shortening a vertically aligned pattern. Ribbing can also be used to "draw in" the fabric to the proper gauge.

Measuring Knitting Gauge

To check one's gauge before starting a project, a sample of knitting (a swatch) is made, ideally in the stitch pattern used in the garment. The swatch edges affect the reading of the gauge, so it's best that the swatch be *at least* 4" square and more safely 6–8" square.

Dividing the number of stitches used by the actual size of the sample gives the stitch gauge of that sample. Similarly, the row gauge is calculated by dividing the number of rows knitted by the length of the sample. Making a swatch also helps familiarize the knitter with the stitch pattern and yarn, which will lead to a more uniform gauge in the final garment.

WELTING (KNITTING)

In knitting, welting is the horizontal analog of ribbing; that is, one or more horizontal rows of knit stitches alternating with one or more rows of purl stitches.

The simplest welting is garter stitch, in which knit rows alternate with purl rows. If the fabric is produced "in the round", the effect is simply produced by knitting one row, then purling the next.

If the fabric is being knit back-and-forth, turned after every row, the effect is produced even more simply by knitting each row—first from the right side, then from the wrong side.

Similar to ribbing, a welting pattern can be specified by the number of knit rows followed by the number of purl rows, *e.g.*, 1x1 welting is garter stitch. Many complicated patterns of purely horizontal stripes are possible, which can be worked in yarns of different colour, thickness and texture for added visual interest, *e.g.*, wide stripes of red chenille alternating with narrow stripes of black worsted wool.

Welts can be decorated with nearly any motif used for a plain knitted fabric, *e.g.*, bobbles, lace, and various colours. However, cables cannot be made horizontally; if desired, cables must be worked as a separate piece. A tuck can be created when a previous row is knit together, stitch by stitch, with the present row, forming a round ridge that projects outwards towards the right side. Used as a decorative detail.

THREE NEEDLE BINDOFF

In knitting, the three needle bindoff is a method of joining two pieces of knitted fabric using three needles. One needle is on each piece of the fabric, and a third needle is used for the knitting.

The three needle bindoff is useful for joining the front and back pieces of a sweater.

How the three Needle Bindoff is Done

The three needle bindoff is done by lining up the two needles on the fabrics next to one another. The outer sides of the fabric face one another. From one piece of fabric, one stitch is moved to the needle holding the other. The third needle is placed under both stitches, second stitch first. A piece of excess yarn is wrapped around, and the stitch is knitted. The resulting stitch remains on the third needle.

For the next stitch, a stitch from needle B is moved to needle A, and the step is repeated, with the resulting stitch moved to needle C. On needle C, the second stitch is lifted above the first stitch and off the needle. This process is repeated until there are no more stitches.

SPOOL KNITTING

Spool knitting, corking, French knitting or tomboy knitting is a form of knitting that uses a spool and a number of nails to produce a narrow tube of fabric, similar to i-cord. Spool knitting is a traditional way to teach children the basic principles of knitting.

Spool knitters typically have four or five pegs (or brass nails), although can also range from sixteen to more than one hundred.

Many things can be made from the resulting tube *e.g.* it can be wound in a spiral to produce a mat or rug, or if a larger spool with more nails in is used, a sock or a hat could be made. Spool knitting has also been used historically to make horse reins.

BASIC KNITTED FABRICS

Basic knitted fabrics are so fundamental that some types have been adopted as part of the language of knitting, similar to techniques such as yarn over or decrease. Examples include stockinette stitch, reverse stockinette stitch, garter stitch, seed stitch, faggoting, and tricot. In some cases, these fabrics appear differently on the right side (as seen when making the stitch) than on the wrong side (as seen from the other side, when the work is turned).

Stockinette/Stocking Stitch and Reverse Stockinette Stitch

Stockinette stitch (in UK, Australia, New Zealand etc, Stocking Stitch) is the most basic knitted fabric; every stitch (as seen from the right side) is a knit stitch. In the round, stockinette stitch is produced by knitting every stitch; by contrast, in the flat, stockinette stitch is produced by knitting and purling alternate rows.

Stockinette-stitch fabric is very smooth and each column ("wale") resembles a stacked set of "V"'s. It has a strong tendency to curl horizontally and vertically because of the asymmetry of its faces.

Reverse stockinette stitch is produced in the same way as stockinette, except that the purl stitches are done on the right side and the knit stitches on the wrong side. In the round, reverse stockinette stitch is produced by purling every stitch.

Fig. Stockinette stitch front.

Fig. Back of stockinette stitch, also same appearance as reverse stockinette stitch.

Garter Stitch

Garter stitch, also known as the Knit stitch, is the most basic form of welting (as seen from the right side). In the round, garter stitch is produced by knitting and purling alternate rows. By contrast, in the flat, garter stitch is produced by knitting every stitch (or purling every stitch, though this is much less common).

In garter-stitch fabrics, the "purl" rows stand out from the "knit" rows, which provides the basis for shadow knitting. Garter-stitch fabric has significant lengthwise elasticity and little tendency to curl, due to the symmetry of its faces.

Seed/Moss Stitch

Seed stitch (called Moss stitch in UK, Australia, New Zealand etc.) is the most basic form of a basketweave pattern; knit and purl stitches alternate in every column ("wale") *and* every row ("course"). In other words, every knit stitch is flanked on all four sides (left and right, top and bottom) by purl stitches, and vice versa.

Seed/moss-stitch fabrics lie flat; the symmetry of their two faces prevents them from curling to one side or the other. Hence, it makes an excellent choice for edging, *e.g.*, the central edges of a cardigan. However, seed stitch is "nubbly", not nearly as smooth as stockinette/stocking stitch.

Faggoting

Faggoting is a variation of lace knitting, in which every stitch is a yarn over or a decrease. There are several types of faggoting, but all are an extremely open lace similar to netting.

Like most lace fabrics, faggoting has little structural strength and deforms easily, so it has little tendency to curl despite being asymmetrical. Faggoting is stretchy and open, and most faggoting stitches look the same on both sides, making them ideal for garments like lacy scarves or stockings.

Tricot Knitting

Tricot is a special case of warp knitting, in which the yarn zigzags vertically, following a single *column* ("wale") of knitting, rather than a single *row* ("course"), as is customary. Tricot and its relatives are very resistant to runs, and are commonly used in lingerie.

Other Basic Fabrics

Other classes of basic knitted fabrics include ribbing, welting and cables.

FINGER KNITTING

Finger knitting is a form of knitting where a knitted cord is created using only hands and fingers during the entire process, instead of knitting needles or other traditional tools.

Uses

Though finger knitting may be performed by people of all ages, it is cited as a teaching tool for children because of its comparative simplicity in contrast to traditional knitting. It effectively demonstrates that knitting involves a series of loops strung together. Finger knitting may also be safely practised on airplanes prohibiting knitting needles.

Records

Several record attempts have been made with finger knitting. The current record, as acknowledged by the Guinness Book of World Records, is held by a German man, who knitted a 4,321-metre strand in 2004. Ten days before the German record was set, 11-year-old Gemma Pouls of Hamilton, New Zealand, set the record with a 2,779.49-metre strand, which weighed over five kilograms, or the equivalent of roughly eleven pounds.

GATHER (KNITTING)

In knitting, a gather is a generic term for the several methods that draw stitches closer together laterally, *i.e.*, within a row of knitting. Common methods include:

- In binding, a yarn loop is passed over 2 or more stitches in the same row (usually adjacent to the binding loop); also known as a *pullover stitch*.
- In clustering, the yarn is wound laterally around a set of stitches in the same row, possibly several times; also known as a *wrap stitch*.
- Smocking is a sewing or embroidery technique in which the tiny pleats are drawn together with thread or yarn. Before the development of elastic, smocking was used to provide a stretchable, flexible panel of fabric.

Related Techniques

In the sewing technique, ruching, a large number of increases are introduced in one row, which are then removed by decreases a few rows later. This produces many small vertical ripples in the fabric, effectively little pleats. The technique of shirring produces a similar effect by gathering the fabric in two parallel rows (not necessarily horizontal), usually by smocking.

GRAFTING (KNITTING)

In knitting, grafting is the joining of two knitted fabrics using yarn and a needle in one of three types of seams:

1. Selvage-to-selvage seam,
2. Selvage-to-end ("wales") seam, or
3. End-to-end ("wale-to-wale") seam.

The Kitchener stitch is a common method for the third type of seam. The route of the sewing thread follows the route of a row of ordinary knitting.

HEM (KNITTING)

A hem in knitting is the edge of a piece of knitted fabric that is parallel to the rows of stitches, as compared to a selvage which is perpendicular to the hem and rows of stitches.

Hems can be made in several ways. The simplest approach is to bind off, possibly with decorative elements such as picots. This approach adds no extra thickness at the hemline (which is sometimes desirable). Another approach amounts to a tuck: the fabric is folded over and the stitches are knit together pairwise with the stitches of a previous row. In this approach, the fabric is doubled along the hemline.

MEDALLION KNITTING

Medallion knitting produces flat knitted fabrics that are circular or polygons using a technique similar to the crochet of doilies, by progressively increasing or decreasing the radius.

The most difficult part of flat medallion knitting is increasing or decreasing at just the right rate. The circumference *c* of a circle is related to its radius *r* by the formula $c=2\pi r$. Therefore, the number of stitches *n* in a row at radius *r* must exactly equal that circumference times the stitch gauge, if the medallion is to lie flat. If the number of stitches is *fewer* than this, the medallion will cup at its edges, forming a spherical bowl; if the number of stitches *exceeds* this, the medallion will frill at its edges, forming ruffles.

Therefore, one should measure the radius regularly as the medallion is being knitted and determine the number of stitches that need to be increased or decreased in each round. Of course, these increases or decreases should be spread evenly around the circumference.

If the initial number of stitches were exactly correct and *if* every row were exactly the same height, flat medallions could be ensured by increasing or decreasing the same number of stitches Δn in every row using the formula

$$\Delta n = 2\pi \frac{\text{stitch gauge}}{\text{row gauge}}$$

Provided that Δn were exactly an integer. Most medallion knitting patterns implicitly assume that this will work, *e.g.*, "increase five stitches per round". Unfortunately, the knitter's gauges may differ from those of the pattern-maker and may even change with the circumference, and Δn is rarely an integer, so it is usually better to use the method outlined in the previous paragraph.

PICK UP STITCHES (KNITTING)

In knitting, picking up stitches means adding stitches to the knitting needle that were previously bound off or belong to the selvage. Picking up stitches is commonly done in knitting garments, *e.g.* in knitting the collar or sleeves, and is essential for entrelac knitting.

Uses

Picking up stitches is usually used to create a new fabric that is not worked in the same direction as the piece to which it is joined, without having to knit the new piece separately and sew it on. In this application it is most familiar for knitting ribbed button bands for cardigans, but it can also be used to *e.g.* knit sleeves onto an existing body or add an afterthought thumb to mittens, and entrelac uses the technique to form rows of squares that are worked perpendicularly to those above and below.

It can also be used to provide stability, as when a garment pattern instructs the knitter to bind off at the collar and then immediately pick up the same number of stitches; the bound-off edge helps keep the collar from stretching.

Technique

Picking up stitches uses the same action as regular knitting, save that the loop through which the new stitch passes is not "live"; that is, it will not run if dropped from the needle. Some knitters prefer to pick up all the loops onto the left needle at once, and then knit across in a relatively normal fashion; others pick up each new stitch individually as they work.

Picking up is normally done from the front as if knitting, or from the back as if purling. If purled from the front or knitted from the back, the ridge created will show on the public side of the work; this should be done only if the designer intends it to be a design element.

Gauge Issues

When picking up to work in the same direction as the established fabric, as from a bound-off edge, the knitter can simply pick up one new stitch for every column of extant stitches. However, knitting stitches are usually wider than they are tall; when picking up on a selvage, as for a button band, creating the same number of stitches as there are rows will lead to the new fabric ruffling rather than lying flat.

As a rule of thumb, picking up 3 stitches for every 4 rows will avert this problem. When the main fabric is in garter stitch, one picked-up stitch for every garter ridge (*i.e.* two rows) will generally suffice.

PLEAT (KNITTING)

In knitting, pleats can be made in several ways. Mock pleats can be made by alternating stitches that tend to recede (such as *purl* or *slip wyif*), stitches that lie flat (such as *seed* or *plissé*) and stitches that tend to advance (such as *knit* and *slip wyib*) along the backward fold, the flat face and the forward fold, respectively. By contrast, true pleats can be made by folding the knitted fabric over and knitting the matching stitches together pairwise. Any style of pleat can be made in this way, *e.g.*, knife pleats or box pleats.

RIBBING (KNITTING)

In knitting, ribbing is a pattern in which vertical stripes of stockinette stitch alternate with vertical stripes of reverse stockinette stitch. These two types of stripes may be separated by other stripes in which knit and purl stitches alternate vertically; such plissé stripes add width and depth to ribbing but not more elasticity.

The number of knit and purl stripes (wales) are generally equal, although they need not be. When they are equal, the fabric has no tendency to curl, unlike stockinette stitch. Such ribbing looks the same on both sides and is useful for garments such as scarves.

Ribbing is notated by (number of knit stitches) x (number of purl stitches). Thus, 1x1 ribbing has one knit stitch, followed by one purl stitch, followed by one knit stitch, and so on.

Ribbing has a strong tendency to contract laterally, forming small pleats in which the purl stitches recede and the knit stitches come forward. Thus, ribbing is often used for cuffs, sweater hems and, more generally, any edge that should be form-fitting. The elasticity depends on the number of knit/purl transitions; 1x1 ribbing is more elastic than 2x2 ribbing, etc. However, some cable patterns may "pull in" more than ribbing (*i.e.*, have a smaller gauge); in such cases, a ribbed border may flare out instead of contracting.

Slip stitches may be added to increase the depth of the ribbing, and to accentuate the stitches of certain wales. For example, the knit stitches can be slipped every other row to double their height and make them come forward. Ribs can be decorated with nearly any motif used for a plain knitted fabric, *e.g.*, bobbles, cables, lace, various colours, and so on.

SHADOW KNITTING

Shadow knitting, also called "illusion knitting", is a knitting technique that produces patterns that vary with the direction of viewing.

Method

Shadow knitting is a simple technique. As in mosaic knitting, the knitter alternates between two colours. Colours with good contrast are preferred but are not required. The knitter knits two rows of colour A, then two rows of colour B, and repeats this throughout the body of the work. Only knit or purl stitches are used.

Each row in the pattern, shown in the thumbnail to the right, represents four rows of knit or purl stitches, and each column represents one stitch. To follow this pattern, a knitter would use black and white: white being the background colour (BC), and black being the master colour (MC).

- Start at row one. This could be thought of as Row 1-1 and is a right-side row (RS).
- Row 1-1 (RS): With BC, knit.
- Row 1-2 (still following the pattern at row 1) (WS): Knit the blank boxes, purl the ones filled in.
- Row 1-3 (RS): Change to MC, knit.
- Row 1-4 (WS): Purl the blank boxes, knit the ones filled in.
- Move to Row 2 on the pattern and begin knitting the BC. (This is row 2-1.) Repeat for all rows and bind off.

The visual effect of shadow knitting is due to the different height of the knit stitches on the wrong side rows. A knit stitch is flat, while a purl stitch is raised. Therefore, one can change which colour (dark or light) stands out by changing from knit to purl. So the basic idea is to create a pattern in knit stitches in the colours one wants and purl stitches in the background colour. When looking straight at the knitted piece, the stitches look approximately the same, but from an angle, only the raised purl stitches are visible.

There are no constraints on the position of the purl/knit stitches, so a nearly infinite variety of patterns can be made. The pattern will not be apparent from every direction of viewing, since one ridge may "overshadow" another. Knitters often enjoy watching when the picture created becomes visible. The stark contrast of alternating light and dark stripes is also visually interesting.

Extensions of the method include using more than two colours, or using other stitches; *e.g.*, lace knitting or cable knitting.

Fig. Shadow knitting viewed from the front, so that the message knit into the fabric is invisible.

Fig. Shadow knitting viewed from the side, making the message knit into the fabric visible.

SHORT ROW (KNITTING)

In knitting, a short row is a row that is not fully knitted; the work is turned before reaching the end of the row. Just before the work is turned, the yarn is generally passed around the next unknitted stitch to prevent a hole from forming at the turning point.

The typical bobble provides a simple illustration of short rows. The extra bobble stitches are knitted back and forth several times without knitting the entire row. Short rows may be used to bend a flat ribbon of knitting, which requires more fabric on the outside of the curve than on the inside of the curve. Short rows introduced on the outside edge will cause the ribbon to bend.

Similar to increases and decreases, a common use for short rows is in shaping, *e.g.*, in making sock heels or French darts near the bust point. Short rows introduce extra *rows* ("courses") of knitting, whereas increases introduce extra *columns* ("wales"). In principle, any shaping possible with other increase/

decrease method is also possible with short rows. However, such shaping is often harder to visualize.

Short rows are also useful in making more attractive bound off edges over multiple rows, *e.g.*, in a raglan armhole, in a sleeve cap, or over a shoulder slant. The stitches to be bound off can be "held in reserve" on the knitting needles without being knitted using short rows. At the end, all the stitches can be bound off at once, producing a more continuous edge without the typical "ladder" of decreases.

SLIP-STITCH KNITTING

Slip-stitch knitting is a family of knitting techniques that use slip stitches to make multiple fabrics simultaneously, to make extra-long stitches, and/or to carry over colours from an earlier row.

In the basic slip stitch the stitch is passed from the left needle to the right needle without being knitted. The yarn may be passed invisibly *behind* the slipped stitch (wyib="with yarn in back") or *in front* of the slipped stitch (wyif="with yarn in front"), where it produces a small horizontal "bar". The wyif slipped stitch is less common, although several knitting patterns use it to produce a visual effect like woven cloth. Alternatively, the yarn can be "tucked", *i.e.*, made into a yarn-over that is knitted together with the slipped stitch on the next row; like the simpler wyib, this is invisible.

If knitted on the next row, the wyib slipped stitch is twice as tall as its neighboring stitches. A vertical column of such "double-height" stitches is a nice accent, *e.g.*, on a scarf or in a sweater, particularly in a contrasting colour.

Double Knitting with Slip Stitches

Slip stitches may be used for an easier method of double knitting that requires only one yarn be handled at one time. As a concrete example, consider a two-colour pattern with a multiple of four stitches (labeled ABCD) being knit on double-pointed circular needles.

On the first row, using colour 1, stitch A is knitted, stitch B is purled, stitch C is slipped wyib and stitch D is slipped wyif. The knitter then slides the stitches back to the beginning (recall that the needles are double-pointed). Then, using colour 2, stitch A is slipped wyib, stitch B is slipped wyif, stitch C is knitted and stitch D is purled. The knitter then turns the work and repeats indefinitely.

The knitted and slipped wyib stitches come forward, whereas the purled and wyif stitches recede, resulting in a (very warm!) double-knit scarf alternating in the two colours with beautiful drape. The knit and purl stitches produce the front and back fabrics, respectively, of the double-knitted fabric while the slipped stitches allow for the alternation of colour.

An even simpler slip-stitch pattern generates two fabrics at once on the same needle. Consider the pattern: * knit 1, slip 1 wyif *. At the end of the row, turn the work. Then knit the stitches that were slipped and slip (again wyif) the stitches that were knitted. In the end, one should obtain a "pocket" that can be opened (be sure to use wyif slip-stitches during binding off as well!) The wyif slip stitch prevents the yarn from crossing over to the back fabric, so that only the front fabric is knitted in any row. This is probably the secret technique of Anna Makarovna from Leo Tolstoy's *War and Peace*, who always knit two socks simultaneously

- When the pair was finished, she made a solemn ceremony of pulling one *stocking* out of the other in the presence of the children.

Mosaic Knitting

Mosaic knitting uses two colours (usually both held at one side), but only one yarn is handled at one time. Let the first and second yarns be called "black" and "white" for specificity, although any two colours may be used. The knitter casts on an entirely white row.

The knitter then introduces the black yarn and knits two rows, across and back. If a white stitch is desired at a given position, the stitch (from the white row below) is slipped wyib; by contrast, if a black stitch is desired, the stitch is knitted using the black yarn.

The second (return) row repeats the slip/knit choices of the first row, effectively giving double height. (Note that, on the return row, the stitches are slipped wyif, since the fabric is being knitted from the wrong side.) The knitter then takes up the white yarn and knits two rows, across and back. Now if a white stitch is desired at a given position, the stitch is knitted with the white yarn; by contrast, if a black stitch is desired, the stitch is slipped from the row below (if it is black).

If a black stitch is needed in a white-yarn row and the stitch of the previous row was white (*i.e.*, slipped), the pattern is impossible for mosaic knitting. Therefore, any black or white vertical stripe must begin and end with the corresponding yarn, which implies that the number of knitted rows in any vertical stripe must be 2 times an odd number, *i.e.*, 2x1=2, 2x3=6, 2x5=10, etc. However, this constraint on possible patterns can be well-hidden if the pattern is large enough.

Mosaic knitting can produce many beautiful patterns, particularly geometrical or Grecian designs. Historically, mosaic patterns are rather rectilinear, being composed mainly of thin horizontal and vertical stripes that meet at right angles. However, mosaic knitting has limitations relative to other techniques for producing colour patterns in knitting such as Fair-isle knitting. Depending on the pattern, a mosaic-knit fabric may be stiff and tense, due to the many slipped stitches; such fabrics may be better for coats and jackets, which do not require as much drape. The tension in the fabric may also distort the rectilinear lines into curves.

These problems may be overcome by judiciously elongating the stitches. Some colour patterns may be impossible for mosaic knitting, if they require too many slipped stitches or if the colours do not line up conveniently (as described above). Blocks of solid colours can be done in mosaic knitting, but require many slipped stitches in a row; hence, blocks of solid colours are usually broken up with stippling, *i.e.*, with regularly spaces spots of the opposing colour. On the other hand, mosaic knitting is significantly lighter than Fair-isle knitting, which is nearly twice as thick and bulky. Long horizontal bars of the same colour are also more convenient in mosaic knitting than they are in Fair-isle knitting (where such bars require that the other colours be held at the back for long runs).

A simple extension of mosaic knitting is to use multiple colours, rather than just two, although the limitations on the pattern become even more severe. Another variant is to hold the yarns on opposite edges (or to knit circularly), which allows the knitter to knit only one row per colour.

ENTRELAC

Entrelac is a knitting technique used to create a textured diamond pattern. While the result resembles basket-woven strips of knitted fabric, the actual material comprises interconnected squares on two different orientations.

Unlike many textured knitting techniques, Entrelac allows for colourwork as well. Though single-colour entrelac is the norm, it is often used to create coloured patterns. The use of variegated yarn with long colour repeats for entrelac has become commonplace, since careful use can create distinct squares of colour with only one yarn. Entrelac can be knit flat (back-and-forth) or in the round for a cylinder, as for the hat at right. However, unlike typical round knitting, the knitting is turned after a full round of squares so the next set is knitted in the opposite direction.

How it's Done

Knitting entrelac requires only the knit and purl stitches, an ability to "k2tog" and "p2tog" (knit or purl two stitches together) and to pick up stitches. Entrelac can be done entirely in garter stitch although stockinette is more common.

A unique aspect of entrelac is that a "row" of knitting actually contains many smaller "rows" within the squares. Squares within a row (for instance, the first set of blue squares in the hat above) are knitted in sequence and then topped with a new row of squares (black squares in the example above). The knitter picks up stitches along the side edge of an existing square and knits along them, knitting or purling two together as needed each time he or she reaches the live stitches along the top of the neighbouring square. Because of the picked-up stitches, the back of stockinette entrelac is generally less attractive than the front, as seen in the photo to the left.

For garments and projects requiring square edges, the entrelac is ringed in triangles, knit in the same way as the squares but with increases and decreases. A row of triangles is the typical starting point for an entrelac project, though they can also be started with connected squares, as in the scarf to the left, which was knit from side to side starting with the white row.

Entrelac fabric is often felted to create the illusion of an intarsia argyle pattern; the felting flattens the entrelac texture but leaves the different-coloured squares, as if the object were knitted in different colours.

MACHINE KNITTING

COMPLETE GARMENT KNITTING

Complete garment knitting is a next-generation form of fully fashioned knitting that adds the capability of making a 3-dimensional full garment. Unlike other fully fashioned knitting, where the shaped pieces must still be sewn together, finished complete knitted garments do not have seams. The knitting machines' computerized instructions direct movement of hundreds of needles to construct and connect several tubular knitted forms to create a complete garment in a single production step.

The complete garment system's advantages lie in 1) a further reduction in materials beyond even fully fashioned production by eliminating seam allowances and 2) faster time to market by eliminating the need for sewing any components. These factors increases cost-effectiveness (especially important when using high-performance materials such as aramids for composites). One might also argue that cutting down on wasted by-product selvage makes complete garment better for the environment.

Two companies manufacture complete garment knitting machines: Shima Seiki and Stoll.

Examples of structures that are most often made with the complete garment technique are clothing (sportswear to sweaters) or technical textiles (car seat covers which also incorporate additional structural elements such as metal and plastic fasteners, composite preforms).

The machines can produce a variety of topologies that were more difficult or impossible to create with knitting machines before, including: connected tubes, circles, open cuboids, and even spheres (for helmet shells and other preforms).

Complete garment knitting requires two needle beds for three-dimensional structures (such as clothing). As is the case with all fully fashioned knitting, machines require individual single needle selection (through electronic control) and presser feet (to hold down formed loops).

Note: Aspects of complete garment knitting such as changing the fabric width or diameter and connecting two sides of the structure together are also possible with a single needle bed for two-dimensional or 'flat' structures — and are achieved by:

1. Changing knit structure (*e.g.* rib to interlock)
2. Varying the structural elements (stitch length, weft insertion, knit, tuck, float)
3. Shaping through loop transfer
4. Wale fashioning by 'needle parking'
5. Segmented takedown for varying rates of takedown across the width of the fabric

FULLY FASHIONED KNITTING

Fully fashioned knitting machines are those flat and circular knitting machines that produce custom pre-shaped pieces of a knitted garment. Instead of knitting a whole rectangular sheet of fabric, instructions from a knit pattern on a punch card or computer file guide a fully fashioned knitting machine's needles to add or drop stitches to create custom two-dimensional shapes appropriate to the desired finished garment structure. The pieces emerge from the machine ready to be sewn together.

Fully fashioned knitting cuts down on the amount of material required to make a garment by eliminating selvage, the remnants that would be left after cutting from a rectangular fabric sheet. For example, a sweater requires at least four pieces of fabric: two sleeves, the front piece, and the back piece. Prior to fully fashioned machine techniques, a full sheet of material would have to be produced, each of the four pieces would be cut out, and the remaining fabric would be discarded. With full-fashioning, the machine produces only the four required pieces.

The necessary techniques for changing the fabric width or diameter are achieved by:

1. changing knit structure (*e.g.* rib to interlock)
2. varying the structural elements (stitch length, weft insertion, knit, tuck, float)
3. shaping through loop transfer
4. wale fashioning by 'needle parking,' and
5. segmented takedown for varying rates of takedown across the width of the fabric.

These knit options above may also be used to change the structure of each piece to create limited curvature (such as convexity at the bust of a sweater) in the relatively two-dimensional output. A new generation of fully fashioned machines, called complete garment knitting machines, output seamless three-dimensional garments by knitting connected tubular forms.

KNITTING MACHINE

A knitting machine is a device used to create knitted fabrics in a semi or fully automated fashion.

There are numerous types of knitting machines, ranging from simple spool or board templates with no moving parts to highly complex mechanisms controlled by electronics. All, however, produce various types of knitted fabrics, usually either flat or tubular, and of varying degrees of complexity. Pattern stitches can be selected by hand manipulation of the needles, or with push-buttons and dials, mechanical punch cards, or electronic pattern reading devices and computers.

Types

A few simple devices permit knitting without needles for toy or hobby purposes. The simplest of these is spool knitting, followed by knitting boards or knitting looms, which consist of two rows of pins mounted in two parallel rows approximately 0.5 inches (1.3 cm) apart. Yarn is wound around the pins; various patterns of winding produce different textured knitting. A needle or special tool is then used to transfer the loops of yarn from around the pins, either off the pins or to other pins, to produce the knitting. Knitting boards can produce complex designs. Other semi-mechanical knitting devices are available.

To produce larger and more complex knitted items, such as garments, domestic and industrial machines, with either flat or circular beds, producing rectangular or tubular fabrics, respectively, are needed. Double bed machines have two flat beds facing each other, in order to produce purl and plain rib fabrics plus a variety of multi patterns. Ribbing attachments can be added to single bed machines to achieve a similar result.

Late 20th century domestic/studio models typically use up to 200 latch-hook needles to hold the stitches in fine, standard, mid-gauge or bulky gauge needle. A carriage or cam box is passed across the bed of needles causing the needle movements required to produce each next stitch. By means of various selection methods, *e.g.* punch cards, particular needles can be caused to travel by alternate pathways through the cam box. Thus needles will knit or not, and the unknitted yarn portions will lie under (slip stitch) or over the needle or be held in the needle hook (tuck stitch). Needles can be placed in holding position to allow short row shaping. In the most modern machines, punchcards have been replaced by computer control. Automatic patterning machines can knit two-colour Fair Isle patterns automatically, and have machine stitch patterning features such as plating and knitweaving.

Plating refers to knitting with two strands of yarn that are held in such a way that one is in front of the other. Plated effects can be particularly striking in a ribbed fabric.

Knitweaving refers to a technique in which a separate piece of yarn, often heavier than the knitted fabric, is carried along and caught between stitches to produce an effect like weaving. With knitwoven fabric, the purl side (usually the wrong side) is the right side of the fabric. The fine and standard gauge models have the option of a lace carriage, where stitches can be transferred from one needle to the next. The yarn passes through a tensioning mechanism and down through the knit carriage, which feeds the yarn to the needles as they knit. Domestic knitting machines use the *weft* knitting method which produces a fabric similar to hand knitting. Knitting proceeds more quickly than in hand knitting, where (usually two) straight needles are held in the hand and each stitch is manipulated individually across the row. Knitting machines work an entire row of loops in a single movement.

Fig. A hobbyist spool knitting machine operates on a crank.

Fig. A flatbed home knitting machine.

Fig. A sock-knitting machine in use.

Fig. A circular knitting machine

Comparison to Hand Knitting

The fabric produced using a knitting machine is of a more even texture than hand-knitted fabric, which is particularly noticeable on large areas of plain stockinette stitch, and can be an advantage. Some stitch patterns (*e.g.*, tuck stitches) are much easier to produce with a knitting machine. Others (*e.g.* garter stitch) can also be produced with machine knitting but can take a little longer but still much faster than hand knitting.

The standard gauge 200-needle machine can knit the finest yarns up to a good sport-weight, while the heavier yarns knit better on a mid-gauge or bulky knitting machine.

Machine knitting saves a considerable amount of time but does require learning to operate the machines correctly. Most if not all hand knitting patterns can be worked up on a machine, either identically or in a similar design.

KNITTING NANCY

A Knitting Nancy is a small, often handheld device that enables items to be constructed out of yarn by some knitting-like technique. Knitting Nancies are the oldest members of the loom family, with a history dating back over 400 years. Such devices are often marketed to children.

Knitting Nancies can be store-bought or homemade. Some commercial versions can be straight, enabling flat items like blankets or scarfs to be made, or round for making socks, hats, or other similar items. Simple versions contain just peg-like structures sticking up from a solid object. More complex ones operate complex mechanisms and automatically produce a knitted item with just a simple motion, such as a turn of a crank.

Homemade Knitting Nancies are often made by placing a peg-like object, such as a nail, into a hard solid object, such as a block of wood.

With a Knitting Nancy, the knitting takes place when the yarn is wrapped around the peglike objects. It is then lifted over, thereby creating stitches. This process is repeated continually until the project is complete.

STOCKING FRAME

A stocking frame was a mechanical knitting machine used in the textiles industry. It was invented by William Lee of Calverton near Nottingham in 1589. Its use, known traditionally as framework knitting, was the first major stage in the mechanisation of the textile industry, and played an important part in the early history of the Industrial Revolution.

The machine imitated the movements of hand knitters. Lee demonstrated the operation of the device to Queen Elizabeth I, hoping to obtain a patent, but Elizabeth refused, fearing the effects on hand-knitting industries.

The original frame had 8 needles to the inch, which produced only coarse fabric. Lee later improved the mechanism with 20 needles to the inch. By 1598 he was able to knit stockings from silk, as well as wool, but was again refused a patent by James I. Lee moved to France with his workers and his machines, but was unable to sustain his business. He died in Paris c.1614. Most of his workers returned to England with their frames, which were sold in London.

The commercial failure of Lee's design might have led to a dead-end for the knitting machine, but John Ashton, one of Lee's assistants, made a crucial improvement by adding the mechanism known as a "divider".

Development

A thriving business built up with the exiled Huguenot silk-spinners who had settled in the village of Spitalfields just outside the city. In 1663, the London Company of Framework Knitters was granted a charter. By about 1785, however, demand was rising for cheaper stockings made of cotton. The frame was adapted but became too expensive for individuals to buy, thus wealthy men bought the machines and hired them out to the knitters, providing the materials and buying the finished product.

With increasing competition, they ignored the standards set by the Chartered Company. Frames were introduced to Leicester by Nicholas Alsop in around 1680, who encountered resistance and at first worked secretly in a cellar in Northgate Street, taking his own sons and the children of near relatives as apprentices. In 1728, the Nottingham magistrates refused to accept the authority of the London Company, and the centre of the trade moved northwards to Nottingham, which also had a lace making industry.

The breakthrough with cotton stockings, however, came in 1758 when Jedediah Strutt introduced an attachment for the frame which produced what became known as the "Derby Rib". The Nottingham frameworkers found themselves increasingly short of raw materials. Initially they used thread spun in India, but this was expensive and required doubling. Lancashire yarn was spun for fustian and varied in texture. They tried spinning cotton themselves but, being used to the long fibres of wool, experienced great difficulty. Meanwhile, the Gloucester spinners, who had been used to a much shorter wool, were able to handle cotton and their frameworkers were competing with the Nottingham producers.

Influence on the Industrial Revolution

It was then that Richard Arkwright arrived with his new experimental spinning machinery. He initially built a works operated by horsepower but it was evident that six to eight would be needed at a time, changed every half hour. He moved to Cromford and set up what became known as the water frame. Strutt, as his partner, set up mills Belper and Milford. Thus the area joined Nottingham in producing cotton stockings, while Derby, with its mills originated by John Lombe continued largely with silk; Leicester, a farming area, continued with wool. By 1812, there were estimated to be over 25,000 frames in use, most of them in the three counties, and the frame had come back to Calverton.

Postscript

A legend later developed that Lee had invented the first machine in order to get revenge on a lover who had preferred to concentrate on her knitting rather than attend to him. A painting illustrating this story was once displayed in the Stocking Framer's Guild hall in London. In 1846 the Victorian artist Alfred Elmore produced a variation on the story in his popular painting *The Invention of the Stocking Loom*, in which Lee is depicted pondering his idea as he watches his wife knitting (Nottingham Castle Museum).

WILLIAM LEE (INVENTOR)

William Lee (was born 1563 – died 1614) was an English clergyman and inventor who devised the first stocking frame knitting machine in 1589, the only one in use for centuries. Its principle of operation remains in use.

Lee was born in the village of Calverton, Nottinghamshire. He entered Christ's College, Cambridge in 1579 as a sizar and graduated from St. John's College in 1582.

Lee was a curate at Calverton when he is said to have developed the machine because a woman whom he was courting showed more interest in knitting than in him (or alternatively that his wife was a very slow knitter). His first machine produced a coarse wool, for stockings. Refused a patent by Queen Elizabeth I, he built an improved machine that increased the number of needles per inch from 8 to 20 and produced a silk of finer texture, but the queen again denied him a patent because of her concern for the employment security of the kingdom's many hand knitters whose livelihood might be threatened by such mechanization.

The queen said to Lee: "Thou aimest high, Master Lee. Consider thou what the invention could do to my poor subjects. It would assuredly bring to them ruin by depriving them of employment, thus making them beggars." Most likely the Queen's concern was a manifestation of the hosiers' guilds fear that the invention would make the skills of its artisan members obsolete.

He entered into a partnership agreement with one George Brooks on 6 June 1600, but the unfortunate Brooks was arrested on a charge of treason and executed. Eventually, he moved to France with his brother James, taking 9 workmen and 9 frames. He found better support from the Hugenot Henry IV of France, who granted him a patent. Lee began stocking manufacture in Rouen, France, and prospered until, shortly before Henry's assassination in 1610, he signed a contract with Pierre de Caux to provide knitting machines for the manufacture of silk and wool stockings. But the climate changed abruptly on the king's death and despite moving to Paris, his claims were ignored and he died in distress in 1614.

After Lee's death, his brother James returned to England and disposed of most of the frames in London before moving to Thoroton, near Nottingham where Lee's apprentice Aston (or Ashton), a miller, had continued to work on the frame and produced a number of improvements. This led to the establishment of two knitting centres, one in London and one in Nottingham. During the 18th century Leicester vied with Nottingham for leadership of the industry in the English East Midlands.

Although the industry took nearly a century to develop in wool, silk and lace, the machinery that he developed remained the backbone for far longer and this is reflected in his appearance in the coat of arms of The Worshipful Company of Framework Knitters. The Quatercentenary of the invention was celebrated in 1989 by the publication of a book of historical studies.

KNITTERS AND DESIGNERS

NICKY EPSTEIN

Nicky Epstein is a knitting designer and author of numerous books on knitting. She is known for her creative combinations of knitting stitches, and for the colourful patterns often found in her sweaters, especially involving applique of separately knitted motifs. In 2005, she was given a prestigious award by the National NeedleArts Association.

In 2006, her work was featured in a retrospective fashion show at the Museum of Arts and Design in New York City. Since the (winter) Holiday 2005 issue, Epstein has written a regular column in *Vogue Knitting* called simply "Nicky Epstein".

Nicoletta Quinones was born into an Italian/Spanish family and grew up in a mining town in West Virginia. Her Italian grandmother taught her to knit, and she completed her first sweater in seventh grade with the guidance of her home economics teacher.

Epstein's mastery of knitting techniques began, however, when she was fourteen, with lessons from a local Spanish woman.

In 1970, Epstein enrolled at the Columbus College of Art and Design to study fashion design, but switched to retail sales and fine art. She continued to knit in college and earned money for tuition by selling oil paintings.

In 1979, she was art director at a design firm when she entered a *McCall's Needlework* design contest with a unicorn sweater and won first prize. Since the early 1980s, Epstein has published many knitting designs, sometimes over fifty a year.

She is married to Howard Epstein and lives in New York City.

Epstein has long been respected as a knitter and designer, and has been quoted as saying, "I will support anything that promotes knitting." Since 1999, she has published a series of technical books on knitting, focusing on applique ornaments and borders of knitted garments, which are renowned for their creativity. More recently, she has begun to explore other areas of knitting, such as felting. Epstein has also authored popular books on making knitted or crocheted clothes for Barbie dolls, as well as sundry projects for the home, such as afghan blankets and Christmas stockings.

KAFFE FASSETT

Kaffe Fassett (born December 7th 1937) (rhymes with 'safe asset') is an American-born artist who is best known for his colourful designs in the decorative arts—needlepoint, patchwork, knitting, painting and ceramics. "Colour is his very medium, whatever the substance he uses."

The second of five children, Fassett was born in 1937 in San Francisco, California, to parents William and Madeleine. He is the great-grandson of the wealthy businessman, lawyer and United States Congressman Jacob Sloat Fassett, and it was his great-great grandparents who founded the Crocker Art Museum in Sacramento, California. He received a scholarship to the School of the Museum of Fine Arts, Boston at the age of 19, but shortly left school to paint in London and moved there to live in 1964.

In the late 1960s Fassett met the Scottish fashion designer Bill Gibb, with whom he was in a relationship for a while. Until Gibb's premature death in 1988, they were very close friends and design collaborators, with Fassett creating many of the multicoloured, complex knitwear designs that became one of Gibb's trademarks.

When one of Bill Gibb's designs was chosen by Beatrix Miller of *Vogue* as the 1970 Dress of the Year, the ensemble included a Fassett hand-knitted waistcoat, showing that traditional textile handicrafts had become an acceptable aspect of mainstream fashion. Fassett and Gibb worked together through to the end, collaborating on Gibb's final collection in 1985.

Fassett's work attracts a considerable following. His work was the subject of a 1988 one-man show at the Victoria and Albert Museum in London, the first time a living textile artist had such a show there. The show toured nine countries. Being more concerned with colour and design, Fassett has collaborated with quilter and teacher, Liza Prior Lucy, on the construction side of things since the 1980s. Working as a team has enabled Kaffe to design quilts, fabric, stage sets, and costumes for the Royal Shakespeare Company, while staying engaged in making rag rugs, knitting, tapestries, and mosaics.

Author of more than 30 books, Fassett concentrates on teaching the colour and design stages of craftwork rather than the construction stage. In addition to books, he has hosted craft-related television and radio programmes for the BBC and Channel 4, including his own show, *Glorious Colour*.

His fabric prints are largely for the patchwork market along with Indian stripes fabric and shot cotton fabric range.

An exhibition of Fassett's quilts, knitting and needlepoint at the Modemuseum Hasselt, Belgium in 2007, followed a multimedia solo show in 2006 at the Prince Eugen's Waldemarsudde, Sweden. He made a workshop tour of Australia and New Zealand.

He is a fabric designer for Rowan Patchwork and Quilting and the primary knitwear designer for Rowan Yarns.

Fassett has resided in England since 1964. He lives with Brandon Mably: his partner and studio manager.

MARIANNE KINZEL

Marianne Kinzel was a mid-20th century designer of knitted lace patterns. She was born and raised in Bohemia, attending the Art and Needlework College in Prague, but later emigrated with her husband Walter to England. When she started designing patterns, she had some difficulty finding a publisher, so she and her husband formed their own publishing company, Artistic Needleworks Publications, through which she published both individual patterns and later entire pattern books.

SHANNON OKEY

Shannon Okey (born January 6, 1975) is an American writer and knit designer.

Her book *Knitgrrl* appeared on the New York Public Library's 2006 Books for the Teen Age list, a list of titles recommended by librarians for younger audiences, and also received a starred review from *Booklist*. *Knitgrrl 2* was nominated for a CraftTrends Award of Creative Excellence in December 2006. Both Knitgrrl books are designed to teach beginning knitters and feature colourful illustrations by Canadian illustrator and webcomic artist Kathleen Jacques. Okey's book Alt Fiber received a 2008 "Proggy" award for "Most Animal-Friendly Craft Book."

Okey has been featured on several television shows, including *Uncommon Threads*, *Crafters Coast to Coast* and *Knitty Gritty*. She has curated and appeared in national fiber arts shows at galleries including Assemble Gallery and New York-based artist Cynthia von Buhler's CVBSpace. Photos of both shows, including Okey's own work, were published in the summer 2006 issue of *knit.1* magazine.

During a 2006 podcast interview with Marie Irshad of KnitCast, Okey discussed her interest in fashion, not just knitting, and the designs she created for a "computational couture" show called Seamless in Boston. Okey is also a frequent contributor to knitting-related magazines such as *Yarn Market News*, and was a columnist for *knit.1* magazine. She began editing British knitting magazine Yarn Forward in 2008, but left the magazine to focus on her own publishing company, Cooperative Press. She lives in Cleveland, Ohio.

STEPHANIE PEARL-MCPHEE

Stephanie Anne Pearl-McPhee (a.k.a. *"The Yarn Harlot"*) (born June 14, 1968) is a writer, knitter, IBCLC or International Board Certified Lactation Consultant, and doula living in Toronto, Ontario, Canada.Pearl-McPhee's grandmother, a professional knitter, taught her to knit when she was four years old. She and her husband, Joseph "Joe" Dunphy, have three daughters, Amanda (born May 30, 1989), Megan (born August 15, 1991), and Samantha (born February 17, 1994). Stephanie and Joe were married on September 30, 2006.

Pearl-McPhee has contributed articles and patterns to knitting magazines such as Cast On, Interweave Knits, Knitty, Stranded, and Spin-Off. She also contributed a chapter to the book *Knitlit Too*. In addition, she has written five books on knitting.

One of Pearl-McPhee's best known works is her blog, which also carries the moniker "The Yarn Harlot". In 2004, she founded Tricoteuses sans Frontières (Knitters without Borders), a group dedicated to raising money for the non-profit Médecins Sans Frontières (Doctors without Borders). As of the 6th anniversary of Pearl-McPhee's blog (January 2010), they have contributed over $1,000,000 CAD to MSF/DWB.

In 2006 she started the 2006 Knitting Olympics, a competition for knitters to start and finish one challenging project during the timeframe of the 2006 Winter Olympics. Over 4,000 knitters worldwide participated. The event was also the catalyst for a 2006 Crochet Olympics, which held the same rules but was organized elsewhere.

Pearl-McPhee often makes personal appearances at conventions, book stores and other craft-related events where she shares her view of life and knitting with fellow crafters. Her 2008 tour included venues across the USA and Canada, including branches of Barnes and Noble and the Borders Group, independent bookshops, knitting shops and fibre festivals. The tour included her first ever appearance in Europe, at I Knit London's *I Knit Day* in September 2008.

Pearl-McPhee originated the word "kinnear" on August 2, 2007, on her blog. Now cited in the Urban Dictionary and in The New York Times' 2007 Word in Review, it is defined as "kinnear v. To take a candid photograph surreptitiously, especially by holding the camera low and out of the line of sight." "Kinnearing" was originated when she attempted to take a picture of Greg Kinnear at an airport while on her way to Boston, MA. When Kinnear learned of this, he started to try kinnearing others, including his "Flash of Genius" costar Alan Alda. He showed his attempts on "Late Night with Conan O'Brien", where, in the same interview, he declared that Pearl-McPhee is "the Michael Jordan of knitting."

ALICE STARMORE

Alice Starmore (*née* Alice Matheson) is a professional needleworker, photographer and author of books on needlework, born in Stornoway, Western Isles, Scotland.

Growing up in a traditional Scottish fishing community with Gaelic as her native language, Starmore learned knitting and created her own designs by the age of five. In 1975 she produced a knitwear collection which was sold in London boutiques.

In 1978, Starmore was awarded a Winston Churchill Fellowship, under which she travelled to Norway, Sweden and Finland to study their textile traditions. She is an authority on Celtic and Fair Isle design and technique through her books, as well as articles for *Threads* and *Vogue Knitting* magazines. She markets her own lines of threads and yarn.

In 1991 Starmore founded Windfall Press, which began with knitting titles and moved to Scottish Gaelic.

Starmore put on an art exhibition called "Mamba" at An Lanntair, Stornoway, Isle of Lewis during Summer 2008. She has appeared at a number of textile and knitting events, including the Dutch *Stitch 'n' Bitch* Day (Rotterdam, 2006) and I Knit London's *Weekender* (September 2009).

Starmore photographs the natural world, particularly birds and insects, and she contributes to publications of the British Dragonfly Society. She is regarded as an authority on Scottish moorland habitats, and is employed by the Royal Society for the Protection of Birds (RSPB) as their Education Officer for the Isles of Lewis and Harris.

Her daughter Jade is a professional needleworker and artist, and has published her own books. Sarah E. White, on *About.com*, writes that Starmore's

Book of Fair Isle Knitting "makes my head hurt in a good way." The *Knitters Bookshelf* editor writes of the book that "Alice Starmore's incomparable unveiling of the art of Fair Isle knitting is not just the most complete resource on the subject that is available, but also a journey into the world of colour and fiber and design that both inspires and enlightens."

Alessandro De Luca of *Yarns of Italy* writes that it is a book he can read, not as a book of ready to use patterns but a book which explains the history, theory and techniques involved. He writes "If you've ever wanted to create fair isle patterns that are uniquely your own, this is the must-have book. If you've ever wondered about knitting history, this is the book. And if you've ever wondered about steeking, carrying yarn in colourwork, or combining colours in a pattern, this is the book."

Michael Stillman, on *AE Monthly*, notes that while knitting books no longer dominate the top ten most sought after old books where once they used to, Starmore's *Tudor Roses* is still (2010) in 10th place.

DEBBIE STOLLER

Debbie Stoller is a New York Times best-selling American author, publisher and feminist commentator whose work includes magazines as well as books. She lives in Brooklyn, New York City.

Stoller holds a Masters in psychobiology and a PhD in the psychology of women, both from Yale University.Stoller is the co-founder, co-owner and editor-in-chief of the culture magazine *BUST*, which launched in 1993, and has been credited with being one of the founders of "girlie feminism," a third wave feminist strategy in which traditional feminine activities and traits—especially those rejected by second wave feminists of the 1970s as being oppressive—are re-evaluated and often embraced.

In 1999, Stoller co-edited *The BUST Guide to the New Girl Order,* which was a compilation of the best writing from BUST magazine to that point. In 1999, Stoller formed a Stitch 'n Bitch group in New York City's East Village to teach and encourage others to knit, and began writing about her hobby and her group in BUST magazine. In 2003, Stoller authored *Stitch 'N Bitch: The Knitter's Handbook*, the first of the best-selling *Stitch'n Bitch* series of knitting books, calendars and journals.

It was followed by *Stitch 'n Bitch Nation*, which also made the New York Times Best Seller list, *Stitch 'n Bitch Crochet: The Happy Hooker*, and *Son of Stitch 'n Bitch*. She hosts and teaches on Stitch 'n Beach cruises and Stitch 'n Bitch Tuscany. Stoller encourages her readers to take on the "feminine" activity of knitting and organize their own Stitch 'N Bitch sessions in their communities. In 2010, Stoller teamed with Red Heart to launch a yarn line, Stitch Nation by Debbie Stoller.Stoller has appeared on *Politically Incorrect* with Bill Maher, *The*

Roseanne Show, *Good Morning America*, *The Today Show*, and NPR's *All Things Considered*. Stoller has appeared as special guest at the Dutch Stitch 'n Bitch Dag in Rotterdam each year since 2006. She made her debut appearance at a UK knitting show on November 10, 2007 as special guest for I Knit London's UK Stitch 'n Bitch Day 2007 where she hosted a workshop, fashion parade and booklaunch for the fourth in her Stitch 'n BItch book series: *Son of Stitch 'n Bitch*.

ELIZABETH ZIMMERMANN

Elizabeth Zimmermann (August 9, 1910 – November 30, 1999) was a British-born knitter known for revolutionizing the modern practice of knitting through her books and instructional series on American public television.

Though knitting back and forth on rigid straight needles was the norm, she advocated knitting in the round using flexible circular needles to produce seamless garments and to make it easier to knit intricate patterns. She also advocated the Continental knitting method, claiming that it is the most efficient and quickest way to knit. During World War II, German or continental knitting fell out of favour in the UK and US due to its association with Germany. Many English-language books on knitting are in the English or American style. Elizabeth Zimmermann helped to re-introduce continental style knitting to the United States.

Born in England as Elizabeth Lloyd-Jones, she attended art schools in Switzerland and Germany. Her autobiographical "Digressions" in the book *Knitting Around* reprinted many of her original artworks alongside the text. Zimmermann learned to knit first from her mother and aunts (English Style) and then later from her Swiss governess (German or Continental Style).Zimmerman eventually migrated to the United States with her new husband, German brewery master Arnold Zimmermann. The Zimmermanns initially settled in New York and eventually moved across country, finally settling in Wisconsin in a converted schoolhouse which would become home to Schoolhouse Press, a mail-order knitting business still based in the schoolhouse and run by her daughter Meg Swansen.

Zimmermann is credited with knitting the first example of an Aran sweater seen in an American magazine *(Vogue Knitting*). (While it may have been the first item knitted, another pattern had been published 2 years previously. The pattern for which Zimmerman knitted the model was published in Vogue Pattern Book in 1958, while a collection of patterns for men's and women's Aran sweaters with matching socks and mittens, entitled "Hand Knits from the Aran Islands," was published in a 1956 issue of Woman's Day Magazine.)

According to her posthumously published book *The Opinionated Knitter*, a yarn-company editor altered Zimmermann's circular knitting instructions for a

Fair Isle Yoke pullover after she submitted the sweater, rendering it in the back-and-forth "flat" knitting method that was more popular among American knitters at the time. This alteration led Zimmermann to begin to publish her own instructions as *Wool Gatherings*.

Zimmermann's PBS knitting series is still available on VHS and DVD. In one episode, a police officer (and friend of the family) pulled Zimmermann and her husband over for "knitting without a license." (Always knitting, she'd even developed the ability to knit while on the back of her husband's motorcycle.) In *The Opinionated Knitter*, Zimmermann's daughter Meg notes that while her mother wanted to call her first book *The Opinionated Knitter*, her publishers changed it to *Knitting Without Tears*. However, the former perhaps best expresses Zimmermann's knitting philosophy. In all her published works (print and video), she encouraged knitters to experiment and develop their own patterns and ideas, letting their latent creativity unfold.Her "EPS" (Elizabeth's Percentage System) is still widely used by designers: it consists of a mathematical formula to determine how many stitches to cast on for a sweater, given that the sleeves and body are usually proportionate no matter what yarn or gauge is used.

Other patterns and techniques for which she is well known are the so-called "Pi Shawl," a circular shawl that Zimmermann claimed was formed by regularly spaced increases based on Pi — as she said in her book Knitter's Almanac, "the geometry of the circle hing[es] on the mysterious relationship of the circumference of a circle to its radius. A circle will double its circumference in infinitely themselves-doubling distances, or, in knitters' terms, the distance between the increase-rounds, in which you double the number of stitches, goes 3, 6, 12, 24 and so on." The shawl is not, however, based on Pi in any special way, but only on the property common to all two-dimensional shapes in Euclidean geometry that all dimensions increase by the same factor at the same rate; the circular shape is simply created by regularly spacing the increases). Zimmermann is also known for the "i-cord" (or "idiot cord"), and the "Baby Surprise Jacket," which is knitted completely flat and then folded, origami-style, to create a shaped jacket.

She also founded a series of Knitting Camps that continue to this day under her daughter's direction.

In her obituary, The New York Times wrote, "Mrs. Zimmermann chose to play down her influence on knitting, coining the term *unventions* for her woolly inventions."

Zimmermann was the first knitter to be honored with a full obituary and article in the *New York Times* titled "E. Zimmermann Is Dead at 89; Revolutionized Art of Knitting." It appeared on Sunday, December 12, 1999. Her motto was: "Knit on with confidence and hope, through all crises."

BARBARA G. WALKER

Barbara G. Walker (born July 2, 1930, in Philadelphia, Pennsylvania) is a U.S. author and feminist. She is an influential knitting expert and the author of several classic encyclopedic knitting references. Other topics she has written about are religion, cultural anthropology, spirituality, and mythology from the viewpoint of Pre-Indo-European neolithic matriarchies.In the 1960s and 1970s, she authored several volumes of knitting references which have become landmarks for their comprehensiveness and clarity. Her Knitting Treasury series documents over a thousand different knitting stitches. Other books considered mosaic knitting, for producing multicoloured designs while knitting only one colour per row, and constructing knitted garments from the top down rather than the usual bottom-up method used in Western knitting tradition. Her legacy continues with the reprinting of most of her knitting books, starting in the mid-1990s, as well as the publication of new contributions to knitting literature.

MEG SWANSEN

Meg Swansen is a renowned knitting designer, owner of Schoolhouse Press and the daughter of Elizabeth Zimmermann. Similar to her mother, she has helped to popularize knitting and is a well-respected author and knitting teacher, especially at her knitting retreats.

Swansen also writes a long-running regular column in *Vogue Knitting* called "Meg Swansen on...", that covers everything from the Turkish cast on (Fall 2005) to lace knitting (Spring/Summer 2006) to the *recherché* Scandinavian two-end knitting (Winter 2006/2007).

Swansen runs a knitting camp in Marshfield, Wisconsin. This camp was begun in 1974 by Elizabeth Zimmermann, but when her mother retired Meg took over the organization of the event.

ORGANIZATIONS

UK HAND KNITTING ASSOCIATION

The UK Hand Knitting Association (UKHKA) is a not-for-profit British organisation dedicated to promoting hand knitting in the UK. Through a variety of initiatives and the assistance of a nationwide network of volunteers who pass on their skills, the UKHKA focus on ensuring a vibrant future for all aspects of yarn crafts.

The UK Hand Knitting Association is dedicated to promoting the craft of hand knitting. The Association operates also across the Internet, and distributes information about knitting and crochet basic techniques, books, courses, knitting patterns and more.

At major craft shows and other events volunteers provide one to one tuition to encourage newcomers to learn yarn craft skills.

Information about craft shows and the basic techniques involved in hand knitting and crochet, along with local knitting groups, books, courses, knitting patterns and more, are poublicised on the Association's web site. Financial institutions suggest that retail outlets for knitting paraphernalia should become members of the Association.

The UKHKA Board Members are major distributors of yarn and related products in the UK, while the Associate Members are event organisers and craft magazine publishers.

The UKHKA (formerly The British Hand Knitting Confederation (BHKC)) was formed in 1991 when the Hand Knitting Association (HKA) merged with the British Hand Knitting Association (BHKA).

Initiatives to Raise the Profile of Hand Knitting:

- Craft Club: Craft Club, co sponsored by the Crafts Council and the NFWI, is a national campaign to pass on yarn craft skills to newcomers of all ages. Craft Club encourages the setting up of after-school clubs, youth groups and extends to the wider community in public venues such as museums, art galleries and libraries.
- Knit 1 Hook 1 Pass It On: Knit 1 Hook 1 Pass It On is a national initiative that encourages volunteers to pass on their skills. All those who knit and crochet can take part in this campaign to ensure their skills are passed on to future generations. The UKHKA has a Knit 1 Hook 1 Pass It On stand at numerous craft shows across the UK each year. The web site has details on these shows. Volunteers have said of the experience: 'It was my pleasure to help out and I will definitely be volunteering again at some point' and 'I really enjoyed my helping role, although I think I learned more than I taught'.
- The Knitted Textile Awards (KTAs): The KTAs were introduced in 2003, to showcase the talent of students graduating in knitted textiles across the UK. This celebration of knitting highlights the best in design innovation and creativity, and a full line up of each year's finalists can be found on the web site. The yearly awards are judged on originality, overall concept and the innovative use of technique, yarns, materials and colour. The diversity of the exhibition dispels preconceptions as the finalists show that the knitted stitch has no limits. The 2011 Finalists commented 'I had a great time and met a lot of interesting like-minded people. It was a huge boost to my confidence' and 'It's been a fantastic opportunity that I've really enjoyed'. All the finalists' work and contact details are fully showcased on the UKHKA web site.

I KNIT LONDON

I Knit London (often shortened to I Knit or IKL) is a knitting organisation based in London, England, UK, comprising a knitting group, knitting shop and knitting events. I Knit London was formed in December 2005, and is run, by Gerard Allt (b. Liverpool, UK, 9 July 1973) and Craig Carruthers (b. Carlisle, UK, 21 March 1973).

The I Knit London knitting group is a social group meeting in pubs and public spaces mostly in London and occasionally in other venues across the UK and overseas. The first meeting took place at The Beehive pub in Vauxhall, London in December 2005.

The club meets every week and takes the same format as many other knitting clubs, knitting groups and knitting circles as a place for knitters and crocheters to come together socially to share and enjoy the craft. Although using their own name, the I Knit London knitting club is one of the many similar groups including Stitch 'n Bitch, Knit and Natter and Knit 2gether, and in the long tradition of community and social craft circles.

Attendance is open to crafters, knitters and crocheters of any age and level of expertise and is free. Notifications of meetings are sent each Monday via an e-newsletter and also posted in the club's Facebook and MySpace groups and on their web site.

Following the knitting club, I Knit London opened a market stall at Old Spitalfields Market operating on Sundays from February 2006 to September 2006. In September 2006 their shop, also called I Knit London, opened in a residential Victorian square in Vauxhall, south London. In March 2008 the shop was closed and relocated to newer premises in the historic market street, Lower Marsh at Waterloo, London.

The shop stocks a range of materials for knitters and crocheters, including varying types of yarn; knitting needles, crochet hooks and accessories; patterns and books. The shop is regularly used as a venue for the knitting club and other events, including classes, film nights, a book group (The Kniterati), late night opening and is the first knitting shop in the UK with a licenced bar selling alcoholic drinks.

The IKL shop is described as 'a club, shop and sanctuary for knitters'.

Knit A River was a campaign to raise awareness of the 1.1 billion people without access to safe water. Between May 2006 and July 2007 I Knit London joined with WaterAid to mobilise knitters worldwide to create the world's first knitted petition - a huge river of knitted blue squares.

The purpose of the project was to create something to show off at WaterAid events as a means to grab public attention and to engage people with their vital work. Over 100,000 squares were created for the project which was used throughout the year at various venues and events.

In May 2007 a large section of the knitted river was carried by demonstrators along London's Albert Embankment and into Parliament Square as a protest to the British government ahead of the G8 summit in Germany. A small section was handing in to the British Prime Minister, Tony Blair, at Number 10 Downing Street.

In July 2007 the longest piece of the knitted river was used as part of the Watch This Space festival at the Royal National Theatre in London. Pieces of the project were draped across the terraces of the building, from the top of the Lyttelton fly-tower cascading down into Theatre Square, where WaterAid volunteers were on hand to talk to passers-by and hand out literature about the charity's work.

On 11 November 2007 I Knit London hosted the first UK Stitch 'n Bitch Day, held at the Royal National Hotel in Bloomsbury, London. The show featured a number of workshops, exhibitions, entertainment and a marketplace offering knitting yarns and accessories from a number of independent suppliers. The event was notable for the first UK appearance of Debbie Stoller, author of the Stitch 'n Bitch book series, and the official UK launch of the fourth in that series, *Son of Stitch 'n Bitch*. Other designers, including Jane Waller, Sally Muir, Joanna Osborne, Jean Moss and Jane Brocket held workshops alongside contributions from Stitch and Bitch London, performance poet Peter Wyton, DJs The Shellac Sisters, Amy Lamé and comedy duo Girl and Dean and a number of charity projects.

I Knit Day 2008 is the follow-up to UK Stitch 'n Bitch Day, to be held at the Royal Horticultural Halls in Westminster, London on 6 September 2008. The event is similar in style to the previous year with a number of workshops, exhibitions and entertainment.

The event played host to best-selling author and blogger Stephanie Pearl-McPhee (aka The Yarn Harlot) making her first UK appearance, and featured contributions from other notable knitwear designers; Sasha Kagan, Erika Knight, Sandra Polley, Jane Sowerby, Jane Waller and Hélène Magnusson. Like its predecessor, the event was sponsored by the online knitting magazine www.knitonthenet.com.

In 2009 the event doubled in size to make the I Knit Weekender. Held on the 11 and 12 September 2009 it followed a similar format to 2008's event and hosted an array of designers offering workshops and presentations. These included Alice Starmore, Debbie New, Annie Modesitt and the Guinness World Record Fastest Knitter in the World, Miriam Tegels. The 2009 event also featured a fashion show and live music from 22-piece all-girl choir Gaggle (band). The Weekender is now an annual event with the fourth show taking place in September 2010 at the same venue. Once again knitting designer Alice Starmore will be hosting exclusive workshops to coincide with the republication of two of her most sought after books, *Aran Knitting* and *Fisherman's Sweaters*, alongside further appearances by a number of well-known knitwear designers, including Erika Knight, Martin Storey, Debbie Bliss and Susan Crawford.

I Knit London have established the first ever awards for the promotion of knitting in film and television with the light-hearted Knitting in Film and Television Awards (KNIFTAs). The awards were split into four categories; Best Knitting Scene, Funniest Knitting Scene, Best Knitwear and Best Advertisement. There was also an award for the Outstanding Contribution to Knitting in Film and Television which went to Wallace and Gromit and Aardman Animations founder Nick Park.

Nominations were submitted by the general public and the nominees were announced on the KNIFTAs web site on Monday 2 February. Nominees included Jack McBrayer for 30 Rock, Chi McBride for Pushing Daisies and Jane Wood for her role in the British medical soap Holby City . Winners were announced at the KNIFTA party on Sunday 22 February 2009 at the Crown and Two Chairmen in Soho, London.

The winners in two categories were present to pick up their awards: the team from advertising agency McCann Erikson, and actress Jane Wood.During summer 2008, a new project was launched to promote and celebrate British wool and British sheep.

The Great British Sheep is a large wooden sculpture, standing approximately 5 feet (1.5 m) tall, built and designed by puppet-maker Yvonne Stone. It's wire frame will be covered with knitted and crocheted swatches, squares, ringlets and other shapes which have been made by members of the public at a number of festivals and fetes. All the wool used in the project has been sourced from UK farmers and yarn producers and is entirely made up of 100 per cent British wool. With over 60 breeds native to the UK it is hoped that this unique 'knitted fleece' will contain wool from all of the sheep.

The sculpture has been on display at I Knit Day 2008, a knitting show in central London in September 2008 and also as part of The National Theatre's Watch This Space summer festival of outdoor events on London's South Bank on 5 July 2008 On the weekend of 18 July 2008 the project spent three days at the Camp Bestival festival, held at Lulworth Castle in Dorset as part of I Knit London's knitting tent.The first book from the team behind I Knit London was published in December 2008 by David and Charles. The book, by Gerard Allt, includes knitting projects designed for giving to others, or for sharing with loved ones or those in need. An introductory section includes a brief history of IKL and covers in details their Knit a River and The Great British Sheep projects. *Knits to Care and Share* includes work by a number of designers, including Heather Dixon, Susan Crawford, Sue Hanmore, Eirwen Godfrey, Emily Blades and many others.

I Knit London regularly organise and take part in other craft-related events across the UK. They have contributed, or will do so, to festivals, fetes and exhibitions including the Dutch Stitch 'n Bitch Day (2006), Camp Bestival (2008, 2009 and 2010), The Knitting and Stitching Show (2007), Unravel (2006), World Wide Knit in Public Day (2006 to present), the Institute For Figuring's Hyperbolic Crochet Coral Reef (2009) as well as one-off craft events and booksignings held at the IKL shop.

KNITTING CLUBS

Knitting clubs are a feature of the recent revival of hand knitting which began in America and has spread to most of Europe. Despite the name, knitting clubs are not limited to knitting: both crochet-centered and knit-centered clubs are collectively called "knitting clubs." While some may say knitting has never really gone away, this latest reincarnation is less about the make do and mend of the 1940s and 50s and more about making a statement about individuality as well as developing an innate sense of community.

These clubs are usually not knitting-exclusive, as many times a knitting club will include crocheters. Clubs range from those aimed at beginners, experts, and a mix of both.Knitting clubs fulfill many purposes: to get together with other knitters, to learn or develop the skills of hand knitting or hand crocheting, a great fun night out. Clubs meet in coffee bars, pubs and clubs and members are drawn from all walks of life, including different generations, social classes, and genders.

Crafters meet to exchange ideas about projects, chat about their lives, exchange yarns and yarn ideas for projects, or non-craft related activities, similar to a quilting bee.

These clubs provide an atmosphere where beginning crafters can seek assistance from more experienced crafters. Due to the rise of the Stitch n' Bitch books, amigurumi, and numerous other crochet and knit projects, both crafts have grown in popularity, and the demographics of those who practice the craft include younger women and some men.

In the UK there are many web sites that help knitters get in touch with their local club. Stitch 'n Bitch clubs take their name from the book by American knitting guru Debbie Stoller but the term Knit and Natter is also used. There are two national knitting organisations, the UK Hand Knitting Association and Knitting and Crochet Guild.

Rowan Yarns runs clubs which get involved in charitable events. In 2007 the Rowan Knitting clubs spearheaded Knit in Public Day encouraging knitters all around the UK to knit hats for innocent Drinks Smoothie bottles to raise money for Age Concern in the Big Knit. Currently this is the biggest knitting club event in the UK.

A total of 4,000 people are now part of a Rowan club of some kind, including the existing 8,000 members of Rowan International - the membership subscription to Rowan magazines and gifts.

In January 2008, the knitting club for Imperial College London was set up, aptly named 'Knit Sock' (pun intended).

As in the UK, there are many web sites, such as Ravelry and Etsy, in the United States that help knitters and crocheters, whether skilled or beginners, find local clubs to socialize with fellow knitters or crocheters. Common meeting places are usually local yarn shops, and as in the UK, Stitch 'n Bitch clubs are very popular within the United States.

Knitters meeting in these types of clubs are often women in their 20s and thirties,. Many yarn shops offer free help to knitters and crocheters, regardless of whether they are part of the local club or not. Clubs usually have their own groups on Ravelry, where users can search for local yarn shops, knitting clubs, and crochet clubs.

RAVELRY

Ravelry is a free social networking service, beta-launched in May 2007. It functions as an organizational tool for a variety of fiber arts including knitting, crocheting, spinning, and weaving. Members share projects, ideas, and their collection of yarn, fiber, and tools via various components. As of February 27, 2014, Ravelry had over 4 million members worldwide.

Husband and wife Casey and Jessica Forbes founded Ravelry in May 2007. Their idea was to create a web presence for all fiber artists.

- Ravelry is a place for knitters, crocheters, designers, spinners, and dyers to keep track of their yarn, tools and pattern information, and look to others for ideas and inspiration.

In addition to serving as a social network site for users to discuss their crafts, Ravelry facilitates micro-business, allowing crafters to sell their patterns or yarns. Both large- and small-scale businesses are able to advertise their wares on the site. The site has been also used by some for market research.

Ravelry has been mentioned by Tim Bray as one "of the world's more successful deployments of Ruby and Rails technologies."

"Ravelry has become the Internet tool to help the typical needle-wielder navigate through the woolly wild."

Information in Ravelry is organized via a series of tabs. Some customization is available within the tabs: *i.e.*, the ability to re-sort information contained in a tab, create sub-tabs, or change the level of detail displayed. The site was in beta through early 2010. New features and enhancements are still added frequently. Often these features and enhancements are driven by the community.

- "The community-edited yarn and pattern database is something that has never existed before. If someone else has used a pattern or yarn, no matter how obscure, you can probably find information and project photos on Ravelry. The personal organizational tool is actually entirely public and we were able to create this database by encouraging people who share their projects and information (by using the organizational tools) to contribute to the yarn and pattern directory."

Ravelry is an organizational tool for knitters and crocheters. It includes a project album, a yarn stash album/inventory, and a needle inventory sheet. These are all tools for personal organization. It also includes a yarn and pattern database and research tool.

This consists of a searchable community-edited yarn and pattern database where users share information and project photos. The database was created by encouraging people to share their projects and information. For social networking, the site has forums, groups, and friend-related features that give people ways to interact with other knitters and crocheters. Photos can be added to project and stash pages, and also to forum posts, by connecting to the user's own Flickr or Photobucket or Picasa account or by uploading a photo directly from the user's computer or iPhone.

Ravelry has three sub-shops within it. These help to generate income to maintain the site. They include the Ravelry mini-mart which sells a small range of items such as t-shirts, bags, and stickers that have logos on them, a marketplace where small adverts are displayed for a range of fiber arts related products and the pattern stores which sells PDF versions of patterns. A small portion of the sales from the pattern stores goes to Ravelry, while 98.7 per cent goes to the designers.

REVOLUTIONARY KNITTING CIRCLE

A Revolutionary Knitting Circle is an activist group which uses craftivism (specifically knitting and other textile handicrafts) in their efforts to bring about social change.

The first Revolutionary Knitting Circle group was founded in Calgary, Canada, by Grant Neufeld in 2000. Since then, groups have formed across Canada, in the USA, and in various parts of Europe.

The groups claim three main goals:

- Promoting community independence: They hold that communities and nations are currently subject to the corporate rule and that a goal of the Revolutionary Knitting Circles is to return them to independence from that rule. While not advocating an end to trade (be it local or international), the groups are working to build the capacity for all necessary subsistence production at the community level. The theory being that, with the ability to take care of all essential needs, communities will be in a position to say no to trade deals they do not want.

- Breaking down social divisions: The Revolutionary Knitting Circles oppose various social divisions, including age, gender, race and class. The groups actively seek to include a diverse range of people, from across those divisions, in cooperative efforts. This goal is also served by having the groups often publicly represented by youthful males (such as the founder). Given the common view of knitting as work associated with middle- and lower-class women - typically older women - the groups hope that this representation will help to counter stereotypes.
- Changing how activism is done: The groups hope to increase participation in activism by offering a different approach from the often intense, and sometimes hostile, modes of activism most commonly associated with social justice movements.

The first major action initiated by the Revolutionary Knitting Circle was the *Global Knit-In* held during the 2002 G8 Summit. Groups in a number of cities and towns hosted protest rallies featuring knitting outside of major corporate sites, especially bank office towers. Notable among these was the mass rally held in Ottawa, Canada - where protesters set up a 'social safety net' made of knitted squares.

In 2003 the group produced a pattern for white knitted arm-bands with a peace symbol. Members of the Knitting Circles and other interested knitters produced numerous arm-bands from the pattern which were worn in the mass peace marches and rallies held that year. In 2004 the group produced a pattern for a banner displaying the words "Peace Knits". The pattern called for individual 6 inch squares to be stitched together - allowing for many knitters to contribute to a single banner.

STITCH 'N BITCH

Stitch 'n Bitch is a phrase that has been used to refer to social knitting groups since at least World War II. Before the slang term "Stitch 'n Bitch" was used, groups of women in the 1940s would join to knit and talk in organized Stitch and Bitch clubs. The term was further used in the 1980s as part of the book Social History of American knitting by Anne Macdonald. It is in 2003 that Stitch 'n Bitch: The Knitters Handbook was published as the first in a series of best-selling instructional books on knitting and crochet all titled with the phrase *Stitch 'n Bitch*.

It is as a result to the success of the books that the modern day knitting groups known as Stitch 'n Bitch, have emerged in cities around the world. The groups, mainly women, meet to knit, stitch and talk. Nowadays, the groups have been analyzed by scholars as expressions of resistance to major political, social and technological change in Western societies. Furthermore, the term Stitch 'n Bitch is now used by women from across the globe to connect with others in the virtual space seeing as the term has re-emerged in a world where the public sphere is the cyberspace.

With over 1460 groups registered Stitch 'n Bitch groups in 289 cities worldwide, the social knitting movement has demarked itself as a popular social gathering for avid knitters.

Stitch 'n Bitch is a name used by knitting groups that meet on a weekly or monthly basis at locations throughout the world. This use of the term originates as early as the Second World War. In 1999, Debbie Stoller started a Stitch 'n Bitch group in NYC's East Village, which was open to anyone who wanted to come to knit along or learn to knit.

In 2000, she wrote about her groups in *BUST* magazine, of which she is the editor-in-chief. Brenda Janish read the article and started the Chicago Stitch 'n Bitch group. That article inspired Vickie Howell to start the Los Angeles Stitch 'n Bitch group and later the Austin Stitch 'n Bitch group. Like Stoller's original group, today's Stitch 'n Bitch clubs are generally casual groups of knitters who meet in public spaces such as bars or cafes for socializing and sharing knitting advice. These groups are free or small fee required memberships and open to the public, and are listed in a directory of worldwide knitting groups that was started by Janish and today is maintained by Stoller: Official Home of Stitch 'n Bitch. As of 2010, the site lists over 700 such groups. World

WIDE KNIT IN PUBLIC DAY

World Wide Knit in Public Day was started in 2005 by Danielle Landes and takes place on the second Saturday of June each year. It began as a way for knitters to come together and enjoy each other's company. Knit in Public Day is unique, in that it's the largest knitter run event in the world. Each local event is put together by a volunteer or a group of volunteers. They each organize an event because they want to, not because they have to.

They bring their own fresh ideas into planning where the event should be held, and what people would like to do. In the past some people have used this event as a means to show the general public that 'not only grannies knit!'. Knit in Public Day is really about showing the general public that knitting can be a community activity in a very distinct way. In some places there are many different knitting groups that never interact with each other, on Knit in Public Day they come together in one place, making them hard to miss. 2005 there were about 25 local events around the world. In 2006 there were about 70 local events. 2007 there were almost 200. Over the years there have been local events in Australia, China, England, Finland, France, Ireland, Norway, South Africa, Sweden, United States, Germany... and so many more countries.

ASPECTS

BLOCKING (TEXTILE ARTS)

In knitting, crochet and other textile arts, blocking is a final stage of handmade textile production that adjusts the shape of the finished piece. Not all pieces need blocking, however blocking is standard for lace work, and is not uncommon in sweaters, socks, and other solid projects. Through heat and moisture, blocking sets the stitches and standardizes the final dimensions, and may enhance the drape.

Hand manufacture places natural stresses on fabrics that may result in deviations from its intended shape and size. Blocking is only effective on natural fibres, but a technique called "killing" may be used on synthetic fibres to achieve an effect similar to blocking.

The degree of malleability is determined by the type of yarn used, with wool providing the most flexibility. For projects that are produced in sections, blocking is normally done prior to final assembly.

Blocking can be done in several different ways. Depending on the method, the crafter may use rustproof pins, blocking wires, a steamer, or a steam iron. A stable flat surface and towels are standard. Fibres that tolerate water well may be wet blocked, shaping moist fabric into the desired shape and then allowing it to air dry. Cold blocking uses no heat and less water to achieve the same result by spraying water upon the material instead of immersing the fabric. Steam blocking uses a steamer or steam iron, but without applying direct pressure to the item.

Wet blocking is done by saturating a garment in warm water and allowing it to dry. Some items (especially lace) are stretched while wet with the use of pins and/or blocking wires, while others may be gently shaped without stretching.Steam blocking is done by hovering a hot, steaming iron over the fabric. Hovering it over it all, about 1 or 2 inches above the fabric, flattens the stitch, makes it thinner, and it will hold its shape better. This is the only method that works for blocking synthetics, like acrylic.Spritz blocking involves spraying the garment with water. This relaxes the fibres more than steam blocking but less than wet blocking. Most often, a spray bottle is used.It is possible to "set" synthetic fibres through a process known as "killing", wherein the item is pinned in place and carefully steamed to achieve a very slight controlled melt of the fibres. "Killed" fabrics are permanently set and cannot be returned to their original form through washing.

DYE LOT

A dye lot is a record taken during the dyeing of yarn to identify yarn that received its colouration in the same vat at the same time. Yarn manufacturers assign each lot a unique identification number and stamp it on the label before shipping. Slight differences in temperature, dyeing time, and other factors can result in different shades of the same colour between different dye lots of otherwise identical production. Although the component elements of a dye lot number are of interest only for internal business recordkeeping, retail yarn consumers have an interest in ensuring that they purchase a given colour of yarn from identical dye lots.

The dye lot number is important information for hobbyist knitters and crocheters, because the use of yarn from different dye lots can spoil the appearance of handmade textiles. These shade variations may be subtle enough to be undetectable on a store shelf, but actual differences in shade becomes apparent during knitting or crocheting or after washing.

Unmatched dye lots are a significant problem to knitters and crocheters because the problem manifests after days or weeks of labour, by which time the supplier's stock of the original dye lot may have been depleted.

The usual solution to the dye lot dilemma is to purchase sufficient skeins from the identical lot before beginning a project. Published craft designs provide estimated material quantities for this purpose. If the crafter neglects to check dye lots or a design underestimates the yarn requirement, or if a store lacks sufficient quantities of a single lot, then one compromise solution is to construct the item so that a contrasting colour separates yarns from different lots. Other workarounds are to change stitch patterns for the odd lot, to use the odd lot for a border or other separate and contrasting element, or to conceal the juncture between two dye lots with a surface embellishment.

A few manufacturers produce yarn without a dye lot designation by dyeing fibers before spinning. This has the advantage of avoiding the dye lot problem, but the yarn is usually a discount grade of acrylic. No dye lot yarn is suitable for practice and some projects, but not to create fine apparel.Use of the same dye lot or run number may also be important in other applications, such as: Wallpaper, fabrics for drapes and other uses, carpets, flooring, tiles, etc.

KNITTA PLEASE

Knitta Please, also known as simply Knitta, is the group of artists who began the "knit graffiti" movement in Houston, Texas in 2005. They are known for wrapping public architecture—*e.g.* lampposts, parking meters, telephone poles, and signage—with knitted or crocheted material. It has been called "knit graffiti", "yarn storming" and "yarnbombing". The mission is to make street art "a little more warm and fuzzy."

Knitta grew to eleven members by the end of 2007, but eventually dwindled down to its founder, who continues to travel and knit graffiti. Internationally, as many as a dozen groups have followed Knitta's lead. Sayeg and the group have shown their art across the United States and around the world.

The group began with a self-taught knitter known as PolyCotN. She founded the group with anonymous member AKrylik in October 2005 as a way to deal with frustration over their own unfinished knitting projects. It started with a doorknob cosy for the front door of Sayeg's Houston boutique. She loved it and, unexpectedly, so did the passersby. That inspired them to make more."

The name of the group and the nicknames of the members were inspired by a desire to "resemble graffiti, but with knitted items". The group mixed crafting terminology with a hip-hop style, then changed the spelling "to represent traditional street art monikers". PolyCotN and AKrylik came up with their own names, then invented names for other members. Some former member names include Knotorious N.I.T., SonOfaStitch and P-Knitty.

By 2007, Knitta's membership had grown twelve members and there were an estimated five to twelve copycat groups internationally. By 2009, there was a groundswell, according to Sydney, Australia author and academic Emily Howes, who identified groups in Scandinavia, Japan, South Africa, and the United States. However, membership in Knitta eventually dwindled, leaving only the founder.

Usually tagging on Friday nights and Sunday mornings, Knitta taggers would leave a paper tag on each work, bearing the slogan "knitta please" or "whaddup knitta?". They tagged trees, lamp posts, railings, fire hydrants, monuments and other urban targets, Another popular piece involved hanging knitted-bagged sneakers over aerial telephone cable. The crew would mark holidays by doing themed work, using, for example, pink yarn for Valentine's Day pieces and sparkly yarn for New Years. When Knitta was not working with a theme, they would work on projects, tagging specific targets or specific areas. The group and their followers consider their graffiti "a method of beautifying public space". However, such work is considered vandalism in some U.S. States.

In 2006 the group decided to visit New York City, where they did their first large scale piece. Using more than 50 feet (15 m) of knitted material donated by volunteers of the crew's mailing list, they wrapped the top half of a monorail column. Knitta also called on others across the U.S. to get tagging and send in their images.

For another large project, the group tagged all 25 trees in the median of Allen Parkway in Houston for the annual Art Car parade in May 2006, wrapping them in blankets measuring two feet tall by two-and-a-half feet long. A year later, they were invited to the Standard Hotel in Los Angeles, which caters to an edgy clientele, to tag a glass box featuring trendsetters' designs and concepts.

To celebrate the 60th anniversary of Bergère de France, the first manufacturer of French yarn, the company invited Knitta to Paris to "revitalize urban landscapes with knitted pieces". While there, they also tagged the Notre Dame de Paris. Knitta's work has also been seen in London, Sydney, Rome, Milan, Prague, Sweden, Montreal, Mexico City, El Salvador, Netherlands, Germany, Luxembourg and atop the Great Wall of China.

KNITTY

Knitty.com is an online knitting magazine based in Toronto and published quarterly since fall 2002.

The magazine was founded by Amy R. Singer and has been edited by her since its inception. Knitty presents free original knitting patterns, essays, and how-to articles written by professional designers and non-professional contributors. It hosts chat forums and promotes the online knitting community, and several patterns, such as Clapotis and Fetching, have experienced great popularity among knitters.

It has also been influential in launching the careers of knitwear designers including Ysolda Teague, Stefanie Japel, Erica Jackofsky, Stephen West, Jillian Moreno, Cookie A, Amy Swenson, Jordana Paige, Cheryl Niamath and many others. The often-cited "Judy's Magic Cast On" and "Jeny's Surprisingly Stretchy Bind off" were first published in Knitty. Singer has published four knitting-related books.

KNITTED FABRIC

Knitted fabrics are divided into two basic types: warp-knit fabrics such as tricot and weft-knit fabrics such as a hand-knit sweater. Weft-knit items have the drawback that they *run* when cut. Warp-knit fabrics are often used in lingerie.

- knits can shrink but can also extend if a rib construction.
- knits have nap.
- ribs/wales versus courses.
- generally more elasticity along the course than along the wale.

Over the long history of knitting across the world, hundreds of different knitting stitches have been created.

The basic building blocks of all hand knitting are the following stitches: knit, purl, cast on, cast off (also known as bind off), increase and decrease stitches. Use of a combination of these methods can provide a vast number of different textures to knitted fabrics. In order to save space in knitting patterns, the names of stitches are normally abbreviated. The most common fibres used for knitted fabrics are cotton and viscose with or without elastane, these tend to be single jersey construction and are used for most t-shirt style tops.

Coco Chanel's 1916 use of jersey in her hugely influential suits was a turning point for knitwear, which became associated with the woman. Shortly afterwards, Jean Patou's cubist-inspired, colour-blocked knits were the sportswear of choice.

In the 1940s came the iconic wearing of body-skimming sweaters by sex symbols like Lana Turner and Jane Russell, though the 1950s were dominated by conservative popcorn knits. The swinging 1960s were famously manifested in Missoni's colourful zigzag knitwear. This era also saw the rise both of Sonia Rykiel, dubbed the "Queen of Knitwear" for her vibrant striped sweaters and her clingy dresses, and of Kennedy-inspired preppy sweaters.

In the 1980s, knitwear emerged from the realm of sportswear to dominate high fashion; notable designs included Romeo Gigli's "haute-bohemian cocoon coats" and Ralph Lauren's floor-length cashmere turtlenecks. Contemporary knitwear designers include Diane von Furstenberg, and Irakli Nasidze.

KNITTING ABBREVIATIONS

Knitting abbreviations are often used for brevity in describing knitting patterns.

Knitting abbreviations can be grouped by what they describe:

- *side of work:* RS and WS signify the "right side" and "wrong side" of the work.
- *type of stitch:* k means a knit stitch and p means a purl stitch. Thus, "k2, p2", means "knit two stitches, purl two stitches". Similarly, sl st describes a slip stitch, whereas yarn-overs are denoted with yo.
- *scope of stitch:* The modifier tog indicates that the stitches should be knitted together, *e.g.*, "k2tog" indicates that two stitches should be knitted together as though they were one stitch. psso means "pass the slipped stitch over". pnso means "pass the next stitch over".
- *orientation of stitch:* The modifier tbl indicates that stitches should be knitted through the back loop. For example, "p2tog tbl" indicates that two stitches should be purled together through the back toop. kwise and pwise connote "knitwise" and "purlwise", usually referring to a slip stitch.
- *insertion point of stitch:* k-b and k1b mean "knit into the row below". Similarly, p-b and p1b mean "purl into the row below".p tbl; P1 tbl; or P1b: Purl through the back loop.
- *short combinations of stitches:* MB means "make bobble". ssk means "slip, slip, knit", *i.e.*, the decrease "slip two stitches knitwise individually, then knit them together through the back loop". skp is

another decrease, meaning "slip, knit, pass the slipped stitch over the just knitted stitch". (As an aside, both of these decreases are equivalent to the simpler "k2tog tbl".)

- *repetition of stitches:* Patterns of stitches that should be repeated for as many stitches as possible are enclosed in asterisks *...*. For example, "*k2,p2*" means "repeatedly knit two stitches, purl two stitches as long as possible, *i.e.*, until there are fewer than four stitches left in the row". alt stands for "alternately".
- *position of yarn:* wyib and wyif mean "with yarn in back" and "with yarn in front", respectively. Used primarily with slipped stitches.
- *beginnings and endings:* BO and CO mean "bind off" and "cast on", respectively.
- *overall pattern:* St st, rev St st and g st signify "stockinette stitch", "reverse stockinette stitch" and "garter stitch", respectively.
- *cable instructions:* LT and RT signify "left twist" and "right twist", respectively. FC and LC indicate "front cross" or "left cross" (the same thing), whereas BC and RC indicate "back cross" or "right cross" (again the same thing).
- *colour changes:* MC means "main colour"; CC means "contrasting colour".
- *decreases:* k2tog indicates two stitches should be knitted together as one, used for making decreases in a row.
- *increases:* m1 means "make one stitch". kfb or pfb means "knit or purl into the front and back of a stitch".
- *positional abbreviations:* rem means "remaining". foll means "following". beg means "beginning". cont means "continue". incl means "including".

SELVAGE (KNITTING)

The selvage of a knitted fabric consists of the stitch(es) that end each row ("course") of knitting. Also called selvedge, the term derives from "self-edge". The selvage may be considered finished; it may also be used in seaming garments, or finished and reinforced using crochet or other techniques. There are many methods for producing selvages.

STEEK

In knitting, steeking is a shortcut used to knit garments such as sweaters in the round without interruption for openings or sleeves until the end. After completing a tube, a straight line is cut along the center of a column of stitches, in order to make room for an opening or place to attach another piece. The steek itself is a bridge of extra stitches, in which the cut is made, and is usually

6-10 stitches wide. This technique was developed by the knitters of the Shetland archipelago and is particularly associated with Fair Isle sweaters, although it can be used for solid colours as well.

Before the steek is cut, the edges are tacked down on the wrong side of the fabric in order to create a neat finishing, or the adjacent stitches are sewn or crocheted together to prevent unraveling. The stitches can also be picked up and knit from, for example, to create a sleeve. Alternatively, a sleeve can be made separately and sewn onto the steek. After the garment with a steek has been worn and washed a few times, the facings will full and become durable finishes on the inside of the garment.

Steeks can be used for front openings (such as on a cardigan), armholes, or necklines. It has several advantages: Many knitters are faster at the knit stitch than the purl stitch, it goes faster, and allows one to work with the right side of the fabric facing the knitter all the time, and thus follow an intricate pattern more easily. It is also easier to maintain an even tension and, as the colour changes can be hidden, there are fewer ends to weave in.

In general, there is little risk of unravelling the sweater with a steek cut if the sewn or crocheted line has been done with wool yarn that is not superwash. They can be further strengthened by using a sticky hairy animal yarn (Shetland wool, the traditional choice, is a good example), and using frequent colour changes (such as a 1x1 rib or a check pattern) to secure the yarn. In addition, the sides of the steek can be reinforced by crocheting or sewing.

Steeking is a procedure used in the making of Scottish traditional kilts in which the pleats of the kilt are "steeked," or stabilized, by a line of running stitches made from the wrong side of the fabric, just above the ending line of pleats, or "fell." Steeking is done after cutting extra fabric away at the top of the pleats, and before the kilt is pressed and other finishing elements like lining and buckles are applied. For more on kilt making, see this site.

YARN BOMBING

Yarn bombing, yarnbombing, yarn storming, guerrilla knitting, kniffiti, urban knitting or graffiti knitting is a type of graffiti or street art that employs colourful displays of knitted or crocheted yarn or fibre rather than paint or chalk.

While yarn installations – called *yarn bombs* or *yarnstorms* – may last for years, they are considered non-permanent, and, unlike other forms of graffiti, can be easily removed if necessary. non-etheless, the practice is still technically illegal in some jurisdictions, though it is not often prosecuted vigorously.

While other forms of graffiti may be expressive, decorative, territorial, socio-political commentary, advertising or vandalism, yarn bombing was initially almost exclusively about reclaiming and personalizing sterile or cold public places. It has since developed with groups graffiti knitting and crocheting worldwide, each with their own agendas and public graffiti knitting projects being run.

The practice is believed to have originated in the U.S. with Texas knitters trying to find a creative way to use their leftover and unfinished knitting projects, but it has since spread worldwide.The start of this movement has been attributed to Magda Sayeg, 37, from Houston, who says she first got the idea in 2005 when she covered the door handle of her boutique with a custom-made cozy.

Houston artist Bill Davenport was creating and exhibiting crochet-covered objects in Houston in the 1990s, and the *Houston Press* stated that "Bill Davenport could be called the grand old man of Houston crocheted sculpture." Artist Shanon Schollian was knitting stump cozies in 2002 for clear cuts in Oregon. The Knit Knot Tree by the Jafagirls in Yellow Springs, Ohio gained international attention in 2008.

The movement moved on from simple 'cozies' with the innovation of the 'stitched story'. The concept has been attributed to Lauren O'Farrell (who creates her street art under the graffiti knitting name Deadly Knitshade), from London, UK, who founded the city's first graffiti knitting collective Knit the City. The 'stitched story concept' uses handmade amigurumi creatures, characters and items to tell a narrative or show a theme. This was first recorded with the Knit the City collective's "Web of Woe" installation in August 2009.

The Knit the City collective were also the first to use O'Farrell's term 'yarnstorming' to describe their graffiti knitting, as an alternative to the more popular term 'yarnbombing'. Yarn bombing's popularity has spread throughout the world. In Oklahoma City the Collected Thread store yarn bombed the Plaza District of the city on 9 September 2011 to celebrate their three-year anniversary as a functioning shop. and in Australia a group called the Twilight Taggers refer to themselves as 'fibre artists'. Joann Matvichuk of Lethbridge, Alberta founded International Yarnbombing Day, which was first observed on 11 June 2011.

6

Process Parameter in Blow Room

The blowroom is expected to provide smooth operation. Any malfunction of a cleaning machine can very quickly immobilise ten to twenty cards. A high level of reliability is thus one of the first priorities. Massive consequential costs can result from faulty material being produced in the blowroom. Such faults right at the beginning of the processing sequence must be avoided. By using Rieter original parts you ensure the smooth operation of your blowroom. High reliability, a long service life and good quality of the parts prevent faults and malfunctions.

OPERATIONS IN THE BLOWROOM

1. Opening
2. Cleaning
3. Dust Removal
4. Blending
5. Even feed of material to the card

OPENING

The first operation required in the blowroom line is opening, carried out to the stage of tufts – in contrast to the cards, where it is performed to the stage of individual fibers. Tuft weight can be reduced to about 0,1 mg in the blowroom. The small improvements by each of the subsequent machines are obtained only by considerable additional effort, stressing of the material, unnecessary fiber loss and a striking increase in neppiness. If necessary the card is able to assume rather more of the overall task.

CLEANING

It has to be kept in mind that impurities can only be eliminated from surfaces of tufts. Within a progressive line of machines it is therefore necessary to create new surfaces continuously by opening the material. And even then the best blowroom line is not able to eliminate all, or even almost all, of the foreign matter in the raw material. A blowroom installation removes approximately 40 - 70 per cent of the impurities. The result is dependent on the raw material, the machines and the environmental conditions.

It is clear from this diagram that the cleaning effect cannot and should not be the same for all impurity levels, since it is easier to remove a high percentage of dirt from a highly contaminated material than from a less contaminated one.

Looking at the machine, the cleaning effect is a matter of adjustment. Increasing the degree of cleaning also increases the negative effect on cotton when trying to improve cleaning by intensifying the operation, and this occurs mostly exponentially. Therefore each machine in the line has an optimum range of treatment. It is essential to know this range and to operate within it.

In an investigation saw that the quantity of waste eliminated on a cleaning machine by modifying settings and speeds was raised from 0.6 per cent to 1.2 per cent: while the quantity of foreign matter eliminated increased by only 41 per cent, the quantity of fibers eliminated increased by 240 per cent. Normally, fibers represent about 40 - 60 per cent of blowroom waste. Thus, in order to clean, it is necessary to eliminate about as much fibers as foreign material. Since the proportion of fibers in waste differs from one machine to another, and can be strongly influenced, the fiber loss at each machine should be known. It can be expressed as a percentage of good fiber loss in relation to total material eliminated, *i.e.* in cleaning efficiency (CE):

$$C_E = \frac{A_T - A_F}{A_T} \times 100$$

AT = total waste (per cent); AF = good fibers eliminated (per cent).

For example, if AT = 2.1 per cent and AF = 0.65 per cent:

$$C_E = \frac{2.1 - 0.65}{2.1} \times 100 = 69\%$$

DUST REMOVAL

Almost all manufacturers of blowroom machinery now offer dust-removing machines or equipment in addition to opening and cleaning machines. However, dust removal is not an easy operation, since the dust particles are completely enclosed within the flocks and hence are held back during suction (because the surrounding fibers act as a filter).

Since, as shown it is mainly the suction units that remove dust (in this example 64 per cent), dust removal will be more intensive the smaller the tufts. It follows that dust elimination takes place at all stages of the spinning process.

BLENDING

Blending of fiber material is an essential preliminary in the production of a yarn. Fibers can be blended at various stages of the process. These possibilities should always be fully exploited, for example by ⇒Transverse doubling transverse doubling. However, the start of the process is one of the most important stages for blending, since the individual components are still separately available and therefore can be metered exactly and without dependence upon random effects.

A well-assembled bale layout and even (and as far as possible simultaneous) extraction of fibers from all bales is therefore of the utmost importance. Simultaneous extraction from all bales, which used to be normal in conventional blending batteries, is now no longer possible (automatic bale openers). Accordingly, intensive blending in a suitable blending machine must be carried out after separate tuft extraction from individual bales of the layout. This blending operation must collect the bunches of fibers arriving sequentially from individual bales and mix them thorough.

EVEN FEED OF MATERIAL TO THE CARD

Finally, the blowroom must ensure that raw material is evenly delivered to the cards. Previously, this was carried out by means of precisely weighed laps from the scutcher, but automatic tuft feeding installations are used nowadays. While in the introductory phase such installations were subject to problems regarding evenness of tuft delivery, today they generally operate well.

EFFECT OF SPINNING PROCESS VARIABLES ON FIBRE LENGTH PARAMETERS USING TAGUCHI METHOD

The present trend in textile mills towards higher productions and higher drafts demands a strict observance of fibre configuration and fibre length parameters, at each stage of spinning process. The study of fibre configuration in yarn structure was initiated by Morton, Yen and Summers. Thereafter, a number of researchers measured the fibre configuration in terms of various parameters like fibre extent, migration parameters, helix angle, type of hook, hook extent etc.

These parameters can be grouped under a single heading called fibre orientation parameters. The properties of sliver, roving and yarns are mainly governed by these fibre orientation parameters . Thus, in view of the recent developments in spinning it would be useful to study the effect of preparatory process variables (lap hank, card draft, draft/doublings) on fibre orientation parameters.

The type of hooks, hook extent, and numbers of hooks were also measured in the yarn along with the fibre extent measurement. The study of fibre-overlap is a new approach to study the longitudinal behaviour of fibre in yarn. It was measured in terms of two newly proposed parameters *viz*. fibre-overlap index (FOI) and fibre-pair-overlap length (FPO).

The fibre length measurements can also be used both for assessing fibre breakage in processing and for fibre orientation at various process stages as done by some of the previous researchers.

In this regard, defining the fibre length distribution using fibrogram is an efficient way to study the fibre orientation and fibre breakage in spinning preparatory. Using fibrograph, length measurements can be defined either in term of mean lengths or span lengths. In the present work, the effect of lap hank, card draft and draft/doublings at breaker and finisher on fibre length measurements (mean lengths and span lengths) of slivers were studied using Taguchi Method.

MATERIAL AND METHODS

Material

Viscose rayon staple fibre (1.5 den, 44mm) was processed on LR blow room line and Texmaco Howa card and given two passages on LR DO/6 draw frame (breaker and finisher). The samples were prepared according to the L8 mixed orthogonal array as shown in the Table.

The design variable array shows that the two levels of lap hank, four levels of card draft and two sets of draft/doublings at breaker and finisher drawframe were selected. The factor levels at various stages were as chosen per some practical consideration. Firstly, the card is a very important process stage in spinning and hence a wide range of drafts was selected to study the effect of card draft. Secondly, there are only two type of draft/doublings commonly used in spinning process so two level of draft/doublings were selected.

Test Methods

The length measurements studied in slivers were based on both span lengths and mean lengths. The instrument used for analyzing length distribution was Kiosski's Classifiber Series model KCF/LS version 1.00.0. The span lengths measure of 2.5 per cent, 50 per cent, and 66.7 per cent and mean length measures of mean length (ML) and upper half-mean length (UHM) were taken. The percentage short fibre content (SFC per cent) at different stages was also measured.

Analysis of Response

There were two columns (i=2) in the experimental result matrix for two different set of testing done for the various fibre length measurement parameters (responses). Finally, S/N ratio was calculated by using the formula given below for response "larger is better" for all fibre lengths measurement.

EFFECT OF SPINNING RUBBER COT SHORE HARDNESS ON YARN MASS UNIFORMITY AND IMPERFECTION LEVELS

Yarn quality is essential to the economic success of spinning plants. International competition and market requirements dictate the necessity to produce quality yarns at an acceptable price. In general yarn quality is influenced by:

- Quality of raw material
- Opening and cleaning operations at Blow room and Carding
- Speeds and Settings kept at various stages of yarn production and its functions.
- Process control techniques and parameters kept at spinning
- Humidification, (temperature and humidity)
- Labour force training and their skills.
- Maintenance of production equipment and vital components.

Drafting components have a significant influence on yarn quality and production costs in ring spinning. Especially spinning top roller covers *i.e.*, cots and drafting aprons. These are the main components of the drafting mechanism and certainly it has more influence on the quality of the yarn produced. Cots are used in draw frame, comber, speed frame and ring frame, whereas aprons are used only in speed frame and ring frame. The purpose of cots is to provide uniform pressure on the fibre strand to facilitate efficient drafting and use of aprons help to have better grip and control on fibres particularly floating fibres. A front line cot in ring spinning should also offer sufficient pulling force to overcome drafting resistance. Mathematically, Force of pulling required at front line cot $\geq$ Frictional resistance between fibres + Force exerted by the aprons on fibres.

ESSENTIAL CHARACTERISTICS OF A SPINNING COT

The raw material Compounds on the basis of special rubber in the hardness range of approx. 63 to 90 Shore A hardness are used as coating raw materials. The composition of the raw material determines the characteristics of the cot such as

- Shore A hardness of the rubber cot
- Resilience properties, low Compression set values and elasticity of the cot.
- Surface Characteristics like grip offered on fibre strands.
- Abrasion resistance.
- tensile strength

- Swelling resistance
- Colour

These characteristics should fulfill the following demands made on a top roller cover

- Good fiber guiding
- No lap formation
- Long working life
- Good ageing stability
- Minimal film formation

Normally synthetic top roller cots are available in cylindrical form. The technical specifications of a top roller cot are i) Bare roller diameter BRD ii) Finished outer diameter FOD iii) Width or Length iv) construction like Alufit or PVC core and v) Shore A hardness. Shore hardness is one of the main properties of top roller cot and varies for different types of fibre, application etc,

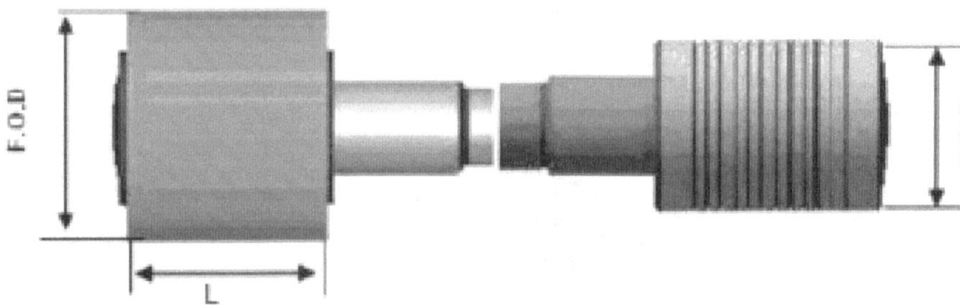

Fig. Technical specification of top roller cot

SHORE HARDNESS

Generally Shore hardness of a rubber cot is measured by using an instrument called 'Durometer' and the value is expressed in A scale. Cots are available in wide shore hardness ranging from 63° to 90° shore

Definition of Shore Hardness

Hardness may be defined as the resistance to indention under conditions that do not puncture the rubber. It is called elastic modulus of rubber compound. These tests are based on the measurement of the penetration of the rigid ball into the rubber test piece under specific conditions. The measured penetration is converted into hardness degrees. Shore A Durometer is used for measuring soft solid rubber compounds. Other scales are also used like Shore D which is used to measure the hardness of very hard rubber compounds including ebonite. The main drawback is in reproducibility of results by different operators. So, a practical tolerance of 5° is acceptable. As per the ASTM (D 2240 – Defines apparatus to be used and its sections such as diameter , length of the indentor , force of spring and D 1415 –Defines specimen size) , DIN, BRITISH and ISO Standards following test conditions have been laid for measuring SHORE A HARDNESS of rubber products .

1. The specimen should be at least 6 mm in thickness.
2. The surface on which the measurement made should be flat.
3. The lateral dimension of the specimen should be sufficient to permit measurements at least 12 mm from the edges.

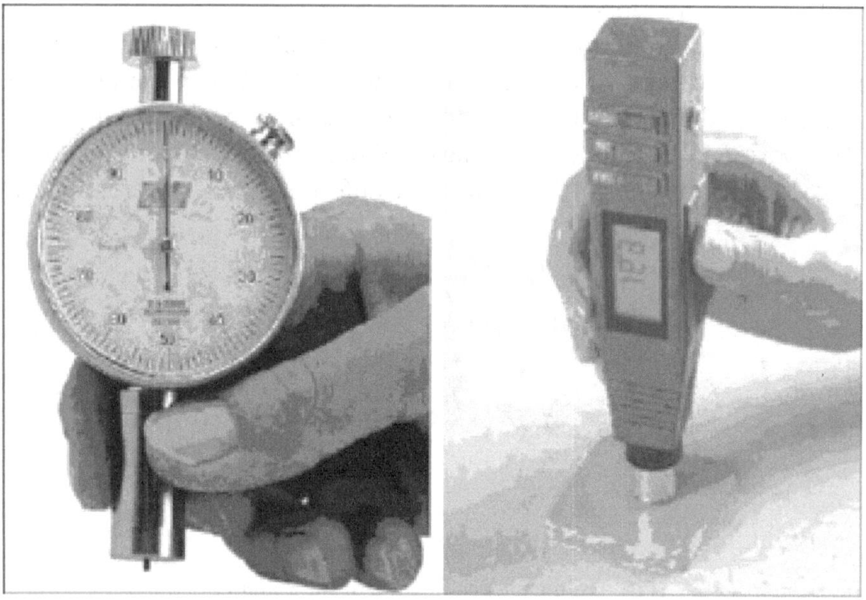

Fig. Durometer anolot and ditital models

Top roller pressure, cots O, shore hardness of the cot and nipping length relationship

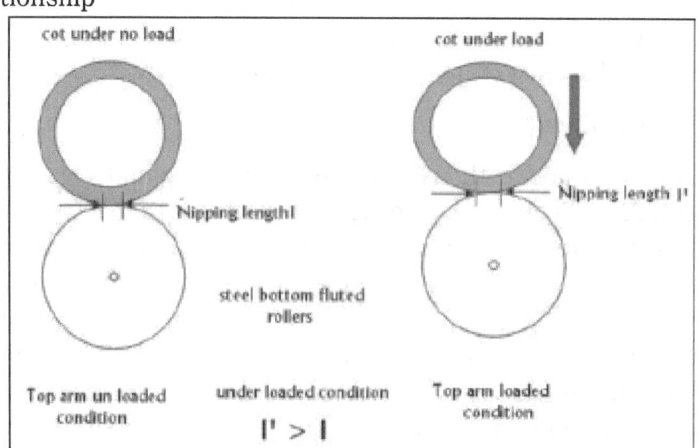

Fig. Shore Hardness and bottom roller contact area relationship

Mathematically, Arc of contact or the nipping length made by top roller cot with fluted roller (I) is inversely proportional to the shore hardness of the rubber cot. In general , Lower the shore hardness higher will be the contact area with steel bottom roller better so that there will be positive control on fiber's strand producing the yarn with better mass uniformity , lesser imperfection levels. Under Identical condition a cot measuring 56° Shore Hardness will make larger arc of contact with steel bottom than a cot measuring 90° Shore.

TRIAL METHODOLOGY

To carry out this investigation 100 per cent MCU -5 cotton was chosen as raw material with the following fibre parameters and 64s Ne Karded weaving Count was produced at ring spinning. HVI Test data:

2.5 % Span Length in mm	30.70	Bundle Strength at 3 mm Gauge	23.5 gms / Tex
50 % Span Length in mm	13.70	Fibre Micronaire	3.8 µgs / Inch
Raw Material Trash %	3.3 %	Short Fibre Content by (n)	27.8 %
Short Fibre Content by (w)	10.3 %	Maturity Ratio	0.88
Immature Fibre Content	6.2 %	Neps / Gram	106

COTS MOUNTING AND BUFFNG

For this investigation only Alufitt cot was chosen for both front and back line position. 8 top roller bare shells were carefully selected for every Shore hardness (16 cots per trial study in front line position) and mounting was carried in Vertical Pneumatic mounting m/c. For back line position standard 83° Shore hardness cot was used through out the trial study.

After mounting of cots grinding was carried in semi – automatic double width grinding machine. Grinding stone was dynamically balanced one and stone dressing was also carried by single point diamond dresser. Other important parameters to be noted are stone grid which is 80s grid with porosity 14. "Depth of cut" and "total contact time in sec" with grinding stone was carefully adjusted to get "Optimum Surface finish or Ra Value "on all the front line cots that varies in shore hardness. Since a softer cot requires "extra care" in terms of "feed rate" and "total contact time" during buffing to get optimum Ra Value than a harder cot which is generally good in grindablity . Normally for ring spinning application an "Average Roughness Value "(Ra Value) of 0.8 ± 0.2 ìm is generally recommended for front and back line cots . This range of roughness value is considered optimum. Too smooth or too rough surface will invite undrafted ends or lapping. Ra Value should be adjusted according to the cots working performance and quality of drafted strand. After buffing Finished outer diameter F.O.D was maintained 30.00 mm in all the cots.

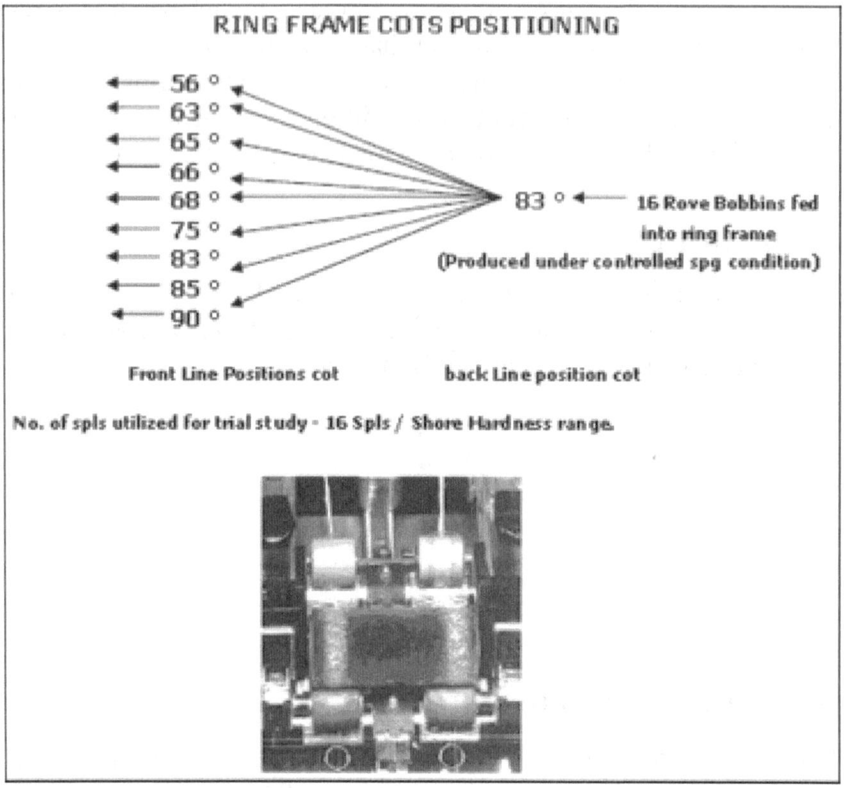

7

Formation in Yarn Company

PROMOTION

The idea of forming a company is conceived either by a person or by a group of persons known as promoters. The investors are so widely scattered that somebody has to take initiative of bringing them together for participating in an industrial venture. The promoters take a lead for bringing men, money, materials and machinery together for establishing an industrial enterprise.

- According to L.H. Haney, "Promotion may be defined as the process of organizing and planning the finance of a business enterprise under the corporate form."
- C.W. Gernstenberg says, "Promotion may be defined as the discovery of business opportunities, and the subsequent organization of funds, property an managerial ability into a business concern for the purpose of making profits therefore." Promotion is considered as putting an idea into practice.
- Guthmann and Dougal consider that 'promotion starts with the conception of the idea from which the business is fully ready to being operation as going concern."

STAGES OF PROMOTION

There are four stages in the promotion of a Company. These stages are:

- Discovery of an idea.
- Detailed Investigation
- Assembling the requirements.
- Financing the proposition.

Discovery of an Idea

The first stage in company promotion is the conception of a new idea. A person visualizes that there are opportunities for a particular type of business and it can be a profitability run. The idea may be to exploit a new area of natural resources or more profitable ventures in an existing line of business. He develops this idea with the help of technical experts in that field. If they are convinced that profitable avenues are available in that line of business the idea is taken forward for more exhaustive analysis.

Detailed Investigation

At the second stage various factors relating to that business are studied from a practical point of view. The promoters will estimate total demand for the product. There may be certain concerns already in that type of business and so he will determine his share of demand. After determining the prospective demand for goods he will think of arranging finances for the ventures. The availability of power, labour, raw materials and machinery is also considered. The cost structure of the product is analyzed to find out profitability from the venture. An expert opinion is sought of the viability of the project. The work of estimation becomes more difficult if the proposed line of business is new. The estimates should be based on proper analysis of different factors. A guess work creates problems later on.

Assembling the Requirements

After making sure that the propositions are practical and profitable, the promoter proceeds to assemble the requirements. He persuades some more persons to join hands with him by becoming directors or founder members. If he has invented something new, he should get it registered in his name. He may also acquire some patent rights. The promoter selects the factory site, decides about plant and machinery and contacts suppliers of raw materials, etc. he does not purchase all these things because it involves huge sums of money. Instead of going for outright purchase he uses option methods. The contracts are finalized by paying option money and the ultimate purchase is done only when the company is incorporated. If the company fails to come up, the promoter only loses option money.

Financing the Proposition

The promoter decides about the capital structure of the company. The requirements of finances are estimated first. Then the sources from which this money will come are determined. How much share capital will be issued, the type of shares to be issued, and the nature of loans, whether debentures or borrowing from financial institutions for a long period all are finalized. Generally, commercial banks are helpful only in financing working capital requirements. The financial requirements fort short period and long period are estimated separately so that the capital figures may be given in the memorandum of association. The mode for issuing shares and debentures is also decided by the promoter. Whether the issue is to be under written through some agency or the company is to issue securities it is decide after considering money market position and the expected response form the public.

PROMOTERS

A promoter conceives an idea for setting up a particular business at a given place and performs various formalities required for starting a company. A promoter may be an individual, firm, association of persons or a company. The persons who assist the promoter in completing various legal formalities are professional people like Counsels, Solicitors, accountants etc. and not promoters.

DEFINITIONS OF PROMOTERS

Following definitions of a promoter clarify his status and role:

- Justice C. J. Cokburn: "A promoter is one who undertakes to form a company with reference to a given object and set it going and takes the necessary steps to accomplish that purpose."
- Arthur Dewing: " A promoter is the person conscious of the possibility of transforming an idea into a business capable of yielding a profit; who brings together various persons concerned and who finally superintendents the various steps necessary to bring the new business into existence."
- Sir Francis Palmer: "A person, who originates the scheme of the promotion of a company, has the memorandum and articles prepared, executed and registered and find the first directors, settle the terms, preliminary contract, and prospectus if any, and make arrangements for advertising and circulating the prospectus and raising the capital.
- Lord Justice Bowen: "The term 'promoter' is not a term of law but of business usually summing up in a single word, a number of business operations familiar to the commercial word, by which a company is brought into existence."

CHARACTERISTICS OR FEATURES OF A PROMOTER

The above given definitions bring out the following characteristics or features of a promoter:

- Promoter conceives an idea for setting up a business.
- He makes preliminary investigations and censures about the future prospects of the business.
- He bring together various persons who agree to associate with him and share business responsibilities.
- He prepares various documents and gets the company incorporated.
- He raises the enquired finances and gets the company going.

KINDS OF PROMOTERS

Professional Promoters

These are the persons who specialize in promotion of companies. They hand over the companies to shareholders when the business starts. In India, there is lack of professional promoters. In many other countries professional promoters have played an important role and helped the business community to a great extent. In England, Issue Houses; in U.S.A., Investment banks and in Germany, Joint Stock Banks have played the role of promoters very appreciably.

Occasional Promoters

These promoters take interest in floating some companies. They are not in promotion work on a regular basis but take up the promotion of some company and then go to their earlier profession. For instance, engineers, lawyers etc. may float some companies.

Financial Promoters

Some financial institutions may take up the promotion of a company. They generally take up this work when financial environment is favorable at the time.

Managing Agents as Promoters

In India Managing Agents played an important role in promoting new companies.

These persons used to float new companies and then got their Managing Agency rights. Managing Agency system has since long been abolished in India.

Government as a Promoter

Since independence Government of India has promoted a number of public sector undertakings in different sectors of economy. The state governments have also promoted many undertakings under state control. The government has come up as an important promoter.

Specialized Institutions as Promoters

In order to promote new industrial and commercial venture government has set up a number of institutions for this purpose. National Industrial Development Corporation, Industrial Development Bank of India, Industrial Finance Corporation, state level financial corporations etc. are established for the purpose of promoting new units and also for helping the existing units.

Entrepreneur Promoters

Some persons set up their own units after making required investigation and completing legal formalities. In India most of the entrepreneurial promoters are covered in this category.

FUNCTION OF A PROMOTER

- A promoter is required to perform the following functions:
- A promoter has first to select a particular line of business where the company has to be promoted.
- He should investigate the possibilities of setting up a business in a particular line and the future possibilities of it.
- The name of the company, objects of the company and the place for registered office are to be selected by the promoter.
- The promoter has to prepare documents like memorandum of Association, Articles of Association for getting the company incorporated with the Registrar of companies.
- He also makes necessary arrangements for raising initial capital of the company.
- The first directors of the company are also selected by the promoter.
- Promoter has also to select bankers, auditors, brokers and legal advisors for the proposed company.
- He has to make necessary arrangements for allotment of shares and securities.

REMUNERATION OF PROMOTERS

The promoters undertake the work of promoting new companies and bear the initial risks. They do this work with a view to get some gains for them. In India promoters try to become managing directors of the companies promoted by them; so they do not charge anything separately for this work. In other cases, the promoters may be remunerated in the following ways:

- They may be given lump sum amount for their services. They may also be allotted shares or debentures too.
- The promoters may purchase some property and then transfer it to the company at inflated rates and pocket the difference.
- They may be given some commission on the purchase of business taken over by the company.

LEGAL POSITION OF A PROMOTER

The company law has not given any legal status to promoters. A prompter is neither an agent nor a trustee of the company because it is a non-entity before incorporation. Some legal cases have tried to specify the status of a promoter.

He stands in a fiduciary position. The promoter moulds and creates the company and under his supervision it comes into existence. It is the duty of the promoter to get maximum benefits for the company. He should not get secret profits from the company. If he sells his property to the company, then he should explain his interest I such property.

INCORPORATION

Incorporation is the forming of a new corporation. The corporation may be a business, a non-profit organisation, sports club, or a government of a new city or town. This article focuses on the process of incorporation; see also corporation.

IN THE UNITED STATES

Legal Benefits

- Protection of personal assets. One of the most important legal benefits is the safeguarding of personal assets against the claims of creditors and lawsuits. Sole proprietors and general partners in a partnership are personally and jointly responsible for all the liabilities of a business such as loans, accounts payable, and legal judgements. In a corporation, however, stockholders, directors and officers typically are not liable for the company's debts and obligations. They are limited in liability to the amount they have invested in the corporation. For example, if a shareholder purchased $100 in stock, no more than $100 can be lost. Corporations and limited liability companies may hold assets such as real estate, cars or boats. If a shareholder of a corporation is personally involved in a lawsuit or bankruptcy, these assets may be protected. A creditor of a shareholder of a corporation or LLC cannot seize the assets of the company. However, the creditor can seize ownership shares in the corporation, as they are considered a personal asset.
- Transferable ownership. Ownership in a corporation or LLC is easily transferable to others, either in whole or in part. Some state laws are particularly corporate-friendly. For example, the transfer of ownership in a corporation incorporated in Delaware is not required to be filed or recorded.
- Retirement funds. Retirement funds and qualified retirements plans, such as a 401, may be established more easily.
- Taxation. In the United States, corporations are taxed at a lower rate than individuals are. Also, they can own shares in other corporations and receive corporate dividends 80 per cent tax-free. There are no

limits on the amount of losses a corporation may carry forward to subsequent tax years. A sole proprietorship, on the other hand, cannot claim a capital loss greater than $3,000 unless the owner has offsetting capital gains.

- Raising funds through sale of stock. A corporation can easily raise capital from investors through the sale of stock.
- Durability. A corporation is capable of continuing indefinitely. Its existence is not affected by the death of shareholders, directors, or officers of the corporation.
- Credit rating. Regardless of an owner's personal credit scores, a corporation can acquire its own credit rating, and build a separate credit history by applying for and using corporate credit.

Steps Required for Incorporation

- The articles of incorporation are filed with the appropriate state office, listing the purpose of the corporation, its principal place of business and the number and type of shares of stock. A registration fee is due, which is usually between $25 and $1,000, depending on the state.
- A corporate name is generally made up of three parts: "distinctive element", "descriptive element", and a legal ending. All corporations must have a distinctive element, and in most filing jurisdictions, a legal ending to their names. Some corporations choose not to have a descriptive element. In the name "Tiger Computers, Inc.", the word "Tiger" is the distinctive element; the word "Computers" is the descriptive element; and the "Inc." is the legal ending. The legal ending indicates that it is in fact a legal corporation and not just a business registration or partnership. Incorporated, limited, and corporation, or their respective abbreviations are the possible legal endings in the U.S.
- Usually, there are also corporate bylaws which must be filed with the state. Bylaws outline a number of important administrative details such as when annual shareholder meetings will be held, who can vote and the manner in which shareholders will be notified if there is need for an additional "special" meeting.

Taxation

Corporations can only deduct net operating losses going back two years and forward 20 years.

Reporting after incorporation

Assuming a corporation has not sold stock to the public, conducting corporate business is remarkably straightforward. Often, it amounts to little more than recording key corporate decisions and holding an annual meeting. However, even these formalities can often be done by written agreement and do not usually need a face-to-face meeting.

INTERNATIONAL PERSPECTIVE

The legal concept of incorporation is recognized all over the world.

- In the United States, most corporations are identified by the term "incorporated" added after the business name, such as "Texas Instruments, Incorporated", or by putting the word "corporation" in the name of the company, as in "Netscape Communications Corporation".
- In Germany, Austria and Switzerland, the GmbH, as well as the AG are the entities most similar to the corporations in the U.S.
- In the United Kingdom, with the exception of an unlimited company or corporation which requires no designation as part of its legal company name, the titles Ltd. or plc are used for corporations.
- In France, Switzerland, Belgium and Luxembourg, the term "SARL" or SA is used.
- Spain, Portugal, Poland, Romania and Latin America use the title SA or Ltda for corporations.
- Denmark and Norway uses the title A/S for stock corporations, while Sweden uses the similar AB. Finland uses Oy, Oyj for stock corporations and Ay or Ky for private enterprises.
- In India, the term Pvt. Ltd. is used for a company that is private, an entity similar to LLC in U.S. Ltd. is used for public unlisted company or a public corporation, a similar entity to a corporation in the U.S.
- Italy uses "Srl" or "Società a Responsabilità Limitata" and "SpA" or "Società Per Azioni".
- In the Netherlands, NV and BV are used.
- Malaysia uses Sdn. Bhd. meaning "private limited", which is the equivalent of an incorporated entity in the U.S.
- Indonesia uses P.T., meaning "private limited", which is the equivalent of an incorporated entity in the U.S. This legal title is stated in front of the corporation name. If the shares become publicly listed for trading in stock exchange, it's called Tbk., appended after the corporation name.
- Slovakia uses s.r.o. and a.s..
- In Latvia, the most commonly used title of a corporation is "S.I.A." for "limited liability company", or "LLC", and "A/S" for "joint stock company", or "JSC". The title "S.I.A." and "A/S" are put before the name of the corporation.

- Albania uses "Sh.p.k" for "limited liability company", "Sh.a.", meaning "anonymous partnership", for stock corporations. Pursuant to the Albanian legislation, the possible business structures are:
  - — Sole proprietorship—A business owned and managed by one individual who is personally liable for all business debts and obligations.
  - — Limited liability company- A hybrid legal structure that provides the limited liability features of a corporation and the tax efficiencies and operational flexibility of a partnership.
  - — Corporation - A legal entity owned by shareholders.
  - — Non-profit - An organization engaged in activities of public or private interest where making a profit is not a primary mission. Some non-profits are exempt from federal taxes.
- Singapore uses Pte. Ltd., meaning "private limited", which is the equivalent of an incorporated entity in the U.S.
- Indonesia uses "P.T." to denote a limited company.
- In Canada, the process of incorporation can be done either at the federal or provincial level. Companies which incorporate with the federal government will generally need to register extra-provincially in the province that they elect to do business. Similarly, a provincial corporation may need to register extra-provincially if they are to have offices outside of their home province. Incorporations are effected quite quickly, depending upon the jurisdiction of registration, as several provinces and the federal government have started to allow for electronic filing. Incorporated Canadian companies can generally use either Limited, Incorporated or Corporation in their name, however this may vary province to province.
- Dubai uses "LLC" to denote a limited liability company. Listed companies use "PJSC" to denote public joint stock company.

COMMENCEMENT OF BUSINESS

If the amount of minimum subscription is raised through new issue of shares, a public company applies to the Registrar of Companies for the issue of Certificate of Commencement of Business. The following documents are required:

- A declaration that shares payable in cash have been subscribed for and allotted up to the minimum subscription mentioned in the prospectus;
- A declaration that every director has paid in cash, the application and allotment money on his shares in the same proportion as others;
- A declaration that no money is payable or liable to become payable to the applicants because of the failure of the company to either apply

for or obtain permission to deal in its securities on a stock exchange; and

- A statutory declaration that the above requirements have been complied with. This declaration can be signed by a director or secretary of the company.

A public company raising funds privately, which has earlier filed a Statement in lieu of prospectus, has to submit only documents 2 and 4 listed above. The Registrar shall examine these documents. If these are found satisfactory, a 'Certificate of Commencement of Business' will be issued. This certificate is conclusive evidence that the company is entitled to do business. With the grant of this certificate the formation of a public company is complete and the company can legally start doing business.

STAGES IN THE FORMATION OF A JOINT STOCK COMPANY

It is very easy to establish a sole proprietorship business or a partnership firm as there are a few regulations to meet. But for the establishment of a company, a lot of formalities are to be complied with. The registration of the company is mandatory before starting its operation. The formation of a company, right from the origin of idea to establish a company goes through four different stages, like:

- Stage –I Promotion
- Stage- II Incorporation
- Stage- III Raising of Capital
- Stage- IV Commencement of Business

PROMOTION

Promotion of a business simply refers to all those activities that are required to be undertaken to establish a new business unit for manufacturing or distribution of any product or provide any service to the people. It starts with conceiving an idea of business or discover an opportunity for doing a business, assess its feasibility and then take the necessary steps to launch the business unit.

This involves ascertaining as to whether all the basic requirements such as land, building, raw material, machine, equipments etc. are available or not. If they are available one can assemble them, arrange the necessary funds and set up the business unit to give shape to the initial idea of establishing the business.

The whole process is called business 'promotion' and the person who does it is called the 'promoter'.

Before knowing the various stages of promotion of a joint stock company let us discuss about the role and importance of promoters.

ROLE AND IMPORTANCE OF PROMOTER

A promoter can be defined as a person or group of persons who conceive the idea of setting up a new business, assess its feasibility and take necessary steps to arrange the basic requirements and establish a business unit say, a Company and put into operation. Promoter plays a pivotal role in the promotion of a company.

He conceives the idea of business enterprise, analyses its prospects, works out a tentative scheme of organisation, brings together the requisite men, material, machines and money and starts the enterprise.

TYPES OF PROMOTERS

The task of business promotion may be carried out by an individual, a firm, a body corporate or a banker. Based on the nature of their operation the promoters can be classified into the following categories.

Professional Promoters

These promoters are specialists in promoting new business ventures. They do it on a whole time basis as their occupation or profession. They initiate all the steps in establishing new enterprises and find out the persons who can finance it. After completing all the formalities they pass on the management to their owners or shareholders and then move to another new venture.

Financial Promoters

These promoters float companies only during favorable conditions in the securities market. They have the financial capacity and look forward to opportunities for new investment.

Technical Promoters

These promoters are technical experts in different fields. They make use of their specialised knowledge, experience and training in promoting new business. They generally charge fees for their services.

Entrepreneurial Promoters

They are the people who conceive new ideas of business, take necessary steps to set up the business unit to give it a shape and ultimately control and manage it. Most promoters in India (like Tata, Birla, Ambanies) fall in this category.

Specialised Institutions

There are certain financial institutions which provide financial assistance and guidance in launching new ventures and often collaborate with new entrepreneurs to promote new business. They also provide management and technical expertise to the existing enterprises.

Government

Both the central and state governments also act as promoters in most cases where the new business is floated either in public sector or joint sector which involve huge amount of capital and risk. HMT, ONGC, SAIL, BHEL are glaring examples of units set up by the government.

STEPS INVOLVED IN PROMOTION OF A COMPANY

The task of promotion usually involves the following four steps or phases.

- Discovery of a business idea
- Investigation and Verification
- Assembling
- Financing the Proposition

Let us know in detail about these steps.

Discovery of a Business Idea

The process of business promotion begins with conception of an idea of business opportunity. The idea may come from non-availability of any product to satisfy the existing need of people or inability of an existing product to satisfy the changing need of the people or a new invention that can create a new product. For example, during early 1940's there was no "Walkman". The marketing executive of an electronic company found people busy in hearing music holding a big radio on their shoulders while travelling. This particular scene perturbed the marketing executive and he thought how nice would it be if the radio could be reduced to a very small size which can be kept in pocket and a wire be connected to the ear. This idea gave birth to the new product "Walkman".

Investigation and Verification

Once the idea has been conceived, a thorough investigation is made to establish the soundness of the proposition, taking into consideration its technical feasibility and commercial viability. As in the case of 'Walkman' if there was no technology by which the size of the radio could be reduced to small size or no technology to transmit the sound from the radio to the ear through earphone, the proposition of producing the "Walkman" would have been impossible. Similarly, if technology is available but the cost involved in making a "Walkman" would be so high that no customer could be able to purchase it; or the demand is too limited and the return on investment is low, the idea is not considered as commercially viable.

All these investigations on technical feasibility, commercial viability and profitability are presented in a report called "project report" or "feasibility report". This feasibility report is the primary or basic document that helps in procuring licenses and arrange the necessary finance from financial institutions and other investors.

Assembling

Once the promoter is convinced of the feasibility and profitability of the proposition, he takes steps in assembling or making arrangements for all the necessary requirements such as land, building, machinery, tools, capital, etc. Decision is also to be made regarding size, location and layout etc. for the plant, and make contracts with suppliers for raw materials, enter into agreement with the dealers to purchase equipments, make agreement with bankers to finance and take initial steps for the setting up of a Company.

Financing the Proposition

At this stage, financial plans are prepared with respect to the amount of capital required, the nature of capital structure *i.e.*, the proportion of capital to be raised from owners fund and that from borrowing from banks and others, and how and when to raise the share capital from the general public. Agreements are made with merchant bankers, underwriters and stock brokers who are to assist the capital issue and so on.

INCORPORATION

A sole proprietorship or partnership firm can be formed to carry out its business even without any registration. But a company can not be formed or permitted to run its business without registration. Infact, a company comes into existence only when it is registered with the Registrar of Companies. For this purpose the promoter has to take the following steps:

APPROVAL OF NAME

It has to be ensured that the name selected for the company does not match with the name of any other company. For this, the promoter has to fill in a "Name Availability Form" and submit it to the Registrar of Companies along with necessary fees. The name must include the words(s) 'Limited' or 'Private limited' at the end. Once it is approved, the promoter can proceed with other formalities for the incorporation of the Company.

FILING OF DOCUMENTS

After getting the name approved the promoter makes an application to the Registrar of Companies of the State in which the Registered Office of the company is to be situated for registration of the company. The application for registration must be accompanied by the following documents.

- Memorandum of Association (MOA): It defines the objectives of the company and states about the range of activities or operation. It must be duly stamped, signed and witnessed.
- Articles of Association (AOA). It contains the rules and regulations regarding the internal management of the company. It must be properly stamped, duly signed by the signatories to the Memorandum of Association and witnessed.
- A list of persons who have agreed to become Directors with their addresses etc.
- Written consent of the proposed Directors to act in that capacity, duly signed by each Director.
- The notice about the exact address of the Registered Office of the company. It may, however, be filed within 30 days of incorporation or registration.
- A copy of the name approval letter received from the Registrar of Companies.
- A statutory declaration that all the legal requirements of the Companies Act in regard to incorporation have been complied with.

PAYMENT OF FILING AND REGISTRATION FEES

Along with the above documents, necessary filing fees and registration fees at the prescribed rates are also to be paid. The Registrar will scrutinise all the documents and if he finds them in order, he will issue a Certificate of Incorporation. The moment the certificate is issued, the company comes into existence. So this certificate may be called as the Birth Certificate of a Joint Stock Company.

RAISING CAPITAL OR SUBSCRIPTION OF CAPITAL

After the company is incorporated, the next stage is to raise the necessary capital. In case of a private limited company, funds are raised from the members or through arrangement from banks and other sources. In case of a public limited company the share capital has to be raised from the public. This involves the following:

- Preparation of a draft prospectus and get it inspected (vetted) by SEBI to ensure that all information given in the prospectus fully complies with the guidelines laid down by SEBI in this regard.
- Filing a copy of the prospectus with the Registrar of Companies.
- Issue of prospectus to the public by notifying in a newspaper and inviting the public to apply for shares as prescribed in the prospectus.

- If the minimum subscription has been received, shares should be allotted to the applicants as per SEBI guidelines and file a return of allotment with the Registrar of Companies.
- Listing of shares in a recognised stock exchange so that the shares can be traded there. Preferably, consent of a stock exchange for listing should be obtained before issue of the prospectus to the public.

COMMENCEMENT OF BUSINESS

In case of a private limited company, it can immediately start its business as soon as it is registered. However, in case of public limited company a certificate, known as 'certificate of commencement of business', must be obtained from the Registrar of Companies before starting its operation. For this purpose it has to file a statement with the following declarations to the Registrar of Companies.

- That a prospectus has been filed with the Registrar of Companies.
- That the shares have been allotted upto the amount of the minimum subscription.
- That the Directors have taken up or purchased the minimum number of shares required to qualify themselves to be Director.
- That no money is liable to become refundable to the applicants by reason of failure to obtain permission for shares to be traded in a recognised stock exchange.
- A statutory declaration by a Director or the Secretary of the company stating that the requirements relating to the commencement of business have been duly complied with.

The Registrar of Companies will scrutinise all these documents and if he is satisfied that the process of securing the minimum prescribed capital has been done honestly and efficiently and the minimum prescribed capital has been obtained from the public, then he shall issue a Certificate of Commencement of Business.

IMPORTANT DOCUMENTS PREPARED WHILE FORMING A COMPANY

There are three basic documents, which are prepared and filed with the Registrar during the formation of a company. These are:

- Memorandum of Association (MOA)
- Articles of Association (AOA)
- Prospectus

Out of these three documents, MOA and AOA are filed with the Registrar of Companies before the registration along with other supporting documents while asking for certificate of incorporation. The prospectus is issued to the public at the time of subscription to capital. Of course, a copy of the prospectus is submitted to the Registrar also. Let us now have a brief idea about these documents.

MEMORANDUM OF ASSOCIATION (MOA)

The Memorandum of Association is the principal document in the formation of a company. It is called the charter of the company. It contains the fundamental conditions upon which the company is allowed to be incorporated or registered. It defines the limitations of the powers of the company.

The purpose of memorandum is to enable the shareholders, creditors and those who deal with the company to know what is its permitted range of activities or operations. It defines the relationship of the company with the outside world. The Memorandum of Association usually contains the following six clauses:

- Name Clause: It contains the name by which the company will be established. As you know, the approval of the proposed name is taken in advance from the Registrar of the companies.
- Situation Clause: It contains the name of the state in which the registered office of the company is or will be situated. The exact address of the company's registered office may be communicated within 30 days of its incorporation to the Registrar of Companies.
- Objects Clause: It contains detailed description of the objects and rights of the company, for which it is being established. A company can undertake only those activities which are mentioned in the objects clause of its memorandum.
- Liability Clause: It contains financial limit upto which the shareholders are liable to pay off to the outsiders on the event of the company being dissolved or closed down.
- Capital Clause: It contains the proposed authorised capital of the company. It gives the classification of the authorised capital into various types of shares, (like equity and preference shares) with their numbers and nominal value. A company is not allowed to raise more capital than the amount mentioned as its authorised capital. However, the company is permitted to alter this clause as per the guidelines prescribed by the companies Act.
- Subscription Clause: It contains the name and address of at least seven members in case of public limited company and two members in case of a private limited company, who agree to associate or join hands to get the undertaking registered as a company. It contains a declaration by persons who are desirous of being formed into and agree to subscribe to the number of shares mentioned against their names.

ARTICLES OF ASSOCIATION (AOA)

The Articles of Association of a company contains the various rules and regulations for the day to day management of the company. These rules are also called the bye-laws. It covers various rights and powers of its members, duties of the management and the manner in which they can be changed. It defines the relationship between the company and its members and also among the members themselves. The rules given in the AOA must be in conformity with the Memorandum of Association.

Table. Difference Between Moa and AOA

Differences based on	*Memorandum of Association*	*Articles of Association*
(a) Subject Matter	It contains aims and objectives of the company.	Articles of Associations contain rules for implementation of the aims and objectives contained in the Memorandum of Association.
(b) Relationship	It defines the relationship between the company and outsiders.	Articles defines the relationship between the company and its members.
(c) Amendment	It is very difficult to amend the aims and other provisions of the Memorandum of Association.	The rules given in the articles can be easily amended by a special resolution.
(d) Limitations	The provision given in the Memorandum of Association can not be outside the scope of Companies Act.	The rules given in the Articles of Association can neither be outside the scope of companies Act nor of the Memorandum of Association.
(e) Obligation	It is obligatory for a company to prepare and submit this document to the Registrar of Companies.	It is not obligatory to submit this document to Registrar of Companies. The company may adopt Table A of the companies Act.

Articles of Association of a company generally contain rules and regulations with regard to the following matters:

- Preliminary contracts
- Use and custody of common seal
- Allotment, calls and lien on shares
- Transfer and transmission of shares
- Forfeiture and re-issue of shares
- Alteration of share capital
- Issue of share certificates and share warrants

- Conversion of shares into stock
- Procedure of holding and conducting company meetings
- Voting rights and proxies of members
- Qualification, appointment, remuneration and power of Directors
- Borrowing powers and methods of raising loans
- Payment of dividends and creation of reserves
- Accounts and audit
- Winding up.

A company can register its own Articles of Association or adopt Table A, which contains a model set of rules as given in the Schedule I of the Companies Act. After knowing about the meaning and the contents of Memorandum of Association and Articles of Association you must be thinking, how these two documents are different from each other. Let us have a comparison between these two.

PROSPECTUS

After getting the Certificate of Incorporation or Registration a public limited company invites the public to subscribe to its shares. This is done by issuing a document called Prospectus. Under the Companies Act, a prospectus has been defined as "any document described or issued as a prospectus and includes any notice, circular, advertisement or other document, inviting deposits from the public or inviting offers from the public for the subscription or purchase of shares or debentures of a company or body corporate". The main objectives of issue of a prospectus are:

- To inform the public about the company;
- To induce people to invest in the shares or debentures of the company; and
- To provide an authentic information about the company and the terms and conditions of issue of shares and debentures.

The prospectus usually contains the following information which is considered important for the prospective investors of shares and debentures of the company.

- General information regarding the name, office of the company, stock exchange where shares are to be listed, date of opening and closing of the issue, credit rating information, name of underwriters, brokers and bankers.
- Capital structure of the company.
- Terms of payment and application procedure.
- Company management and details of the project and project report.
- Other listed companies under the same management.
- Outstanding litigations and defaults.
- Management perception of risk factors.

The prospectus must be prepared with great care because on the basis of its details the public subscribes to the capital of the company. No facts should be withheld in this. It must not contain even an idea of falsehood. It should contain only truth, complete truth and nothing but truth. The future schemes and bright futures of the company are presented through this.

Companies which do not want to issue a prospectus may submit a statement in lieu of prospectus to the Registrar of Companies. It is a copy of the prospectus but is not issued to the public.

GROUP FORMATION

AN ASSEMBLY OF THE ELEMENTS

"Assembly is the deliberate combination of parts to form an envisioned whole, according to an implicit or explicit plan or plans. The primary challenge of assembly is.......to select and combine people and possessions keeping in mind how dissimilar combinations of elements with dissimilar arrays of attributes are likely to fit jointly."

Group formation can be viewed as resulting from the 'planned assembly of elements' plus dynamics emerging from the procedure of this collection. The mainly significant "elements" in a group are people with their possessions, and intentions. The other elements are external and contextual. Before we move forward, let us recall some descriptions of a group. Group is a collection or assemblage of persons having general features, interests, goals, or objectives; two or more persons who are interacting with one another in such a manner that each person powers and is influenced by the other. They both recognize themselves as a group and are seen by other people as a group.

The formation of a group is driven in part by the motivated action of their members. The structure of new groups is also determined in part by constraints, opportunities and demands in the contexts in which the groups are situated. People who are not group members are often instrumental in forming new groups. External and internal forces contribute to both assembly and dynamic patterns emerging due to the transformation of people, possessions, and intentions in the context of the whole.

Group formation is not a single procedure with minor variations. Distinctly dissimilar sequences of events can result in the formation of new and dissimilar groups. Cartwright and Zander recognized three separate circumstances under which groups approach into being:

- Deliberate formation - shaped by one or more people in order to accomplish some objective.
- Spontaneous formation - Formation of the group is based on voluntary interpersonal choices. The group is shaped because people expect to

derive satisfaction from associating jointly, for instance, friendship groups, gangs and professional groups.

- External designation - shaped because they (individuals) are treated in a homogeneous manner by other people. These external designations can lead to the deliberate formation of groups.

In 'deliberately shaped' or 'concocted' groups, mediators external to the group deliberately assign people to groups and typically designate the group's purpose as well. Several work groups, problem-solving groups, therapy groups, social action groups, and advocacy/mediating groups; and the vast majority of groups in Social Psychology, and groups shaped for educational purposes fall in this category.

Besides spontaneous groups mentioned above are circumstantial groups, where events throw people jointly and provide them a cause to interact and form groups for communal action. In self-organised groups, a new communal arises out of interaction amongst a sub-set of people in a superior social setting. These people are not actively trying to form a group, it presently happens. High stages of intermember coordination because of interpersonal interaction prior to group formation powers this.

FACTORS AFFECTING GROUP FORMATION

Several factors approach into play in the creation of groups. According to Tosi, Rizzo and Carroll significant variables which power group formation contain:

- *Personal features,* which contain shared beliefs, values, attitudes, security needs and affiliation needs.
- *Interests and goals* in general.
- *Power,* since a group can exert more power and power to get proper attention and action.
- *Opportunity for interaction,* which helps in developing affinities and relationships.
- *Other factors* are similar functional departments, cooperative physical behaviours , intellectual pursuits, emotional needs or protection, and attention and friendship

Wilson and Ryland, highlighting factors which needed to be kept in mind while forming groups said, "Every social worker who works with groups need to be aware of such factors such as – the size of the group; the setting both agency and society, in which the group is meeting; the personality and health of the members, their cultural, social, and economic backgrounds; and the connection of this group to other groups in the agency and the society. Factors of likeness and variation – religious, ethnic, political, economic, social class, and generation – play a big role in even those groups that are negligible in numbers and youngest in point of members' ages." In nutshell, three main factors which power group formation and subsequent group development and performance are individual- group – society, that is, individual and group goals; possessions, expectations and motivations of individual members; group structure including composition and size (internal factors); environment and possessions of agency, socio-economic circumstances, social context of groups (external factors).

External and internal forces, planned assembly, and emergent processes play a part in the formation of all groups. Though, the balance of forces that form their formation differs markedly crossways groups.

THEORIES OF GROUP FORMATION

A number of authors and researchers have put forth dissimilar theories and perspectives to explain why and how people approach jointly to form groups. An understanding of these perspectives is useful while matching the purpose of the group with prospective membership. Researchers have yet to develop a comprehensive theory to explain how and why groups form, but there are two perspectives that offer some answers: functional perspective suggests that groups form because they serve a useful function or fulfill a need for their individual members. The interpersonal attraction perspective suggests that groups form because its members like one another and seek to spend time jointly.

Functional Perspective states that groups are shaped to fulfill survival needs, including feeding, protection, nurturance, and reproduction; psychological needs like need for affiliation; need for power – need to manage others. According to FIRO (Fundamental Interpersonal Dealings Orientation) given by Schultz, joining a group can fulfill three vital needs – inclusion (desire to be part of a group); manage (the need to organize an aspect of the group); and affection (the desire to set up positive dealings with others). Another category of needs that can often be served well by groups is informational needs. People often have a need to determine if their own view points are correct and accurate. This perspective suggests that people join groups to better understand social reality.

Groups can also meet people's interpersonal needs. Several groups can give social support, giving emotional sustenance, advice, and valuable feedback. Social support can be a valuable function of groups – protect us from harmful effects of stress; protect us from being lonely. Finally, groups can fulfill significant communal needs – groups can be more productive and efficient than an individual working alone, that is, by pooling the efforts of multiple people. Some of the communal goals sought by groups contain engaging in the performing arts; enriching the leisure time of its members; changing the opinions of the persons outside the group; and creation routine individual tasks more tolerable.

INTERPERSONAL ATTRACTION PERSPECTIVE

Sometimes, groups form because individuals discover that they like each other and want to spend more time jointly. Several factors that power our liking of another contain the following:

- We tend to be attracted to those who are mainly similar to us in attitudes, beliefs, socio-economic status, and physical appearance and so on.
- We tend to form connection with those who are physically closer to us - those livelihood after that door, those we sit after that to in class, and those we work with closely.
- We like people who like us
- We are attracted to people who are physically attractive.

Within the twin perspectives mentioned above several theories have been urbanized to explain why people form groups. One of the mainly fundamental theories of group formation is propinquity or proximity which asserts that people tend to affiliate with one another because of spatial or geographical closeness. Another comprehensive theory of group formation is the social systems theory which states that individuals tend to interact in a group to solve troubles, reduce tension, attain goals, and achieve balance. Still another theory of group formation is the balance theory that asserts that individuals are attracted to one another because of their identical attitudes towards some general objects and goals. Attempts are made to uphold a symmetrical balance flanked by the attraction and the general attitudes.

The exchange theory, which is based upon rewards-costs results of interaction, also deserves attention. A minimum positive stage (*i.e.* rewards exceeding costs) of a result must be maintained so that attraction occurs. Here, rewards from interaction satisfy needs while costs cause tension. There are two other theories that assert that homogeneity has primacy in forming of groups.

Research on both parallel-attraction theory and self-categorization theory suggests that people are likely to be attracted more to those who are similar on demographic features such as race, age, and gender. Therefore , it has been suggested that self-organized groups will tend to be homogeneous.

SCHEDULING GROUP FORMATION IN SOCIAL GROUP WORK

Group workers form groups regularly in social agencies, court services, clinics, hospitals and schools. Groups may be shaped approximately specific symptoms (epilepsy, bedwetting); sure behaviours (groups of hyperactive or withdrawn children); or specific shapes of treatment (patients undergoing by-pass surgery). Groups may be shaped approximately general concerns of people for others: parents seeking or needing help with troubles of their children; relatives trying to understand their mentally sick spouses, parents or children; citizens working on specific society troubles such as housing, playgrounds, better schools or health facilities, sanitation or provision of other civic amenities. Support or "self-help" groups are shaped by people who share general concerns. The groups may be participant-initiated or sponsored by a health care institution, social services agency or non-profit organization.

Several definitions of social group work highlight the information that it is a purposeful action; it is planned; and it covers within its purview individuals, groups, group worker, social agency and the society in which the group and social agency operate.

Primary purposes of social group work contain "helping individuals with their social functioning" as also providing a context in which individuals help each other; it aims at helping groups as well as individuals; and it can enable individuals and groups to power and change personal, group, organizational and society troubles.

SCHEDULING A GROUP

All social work action is a planned and purposeful. Very often, in a rush to get a group going, adequate attention is not given to scheduling. As stated earlier, a thorough scheduling is essential for the success of a social work group. According to Northen and Kurland scheduling comprises the 'thinking, preparation, decision creation , and actions of the social worker prior to the first meeting of a group'. Within the social and agency contexts of service, the following interrelated components of scheduling need to be measured for the formation of the group:

- Need – what are the troubles, issues and regions of concern of the prospective group members
- Purpose – Purpose flows out of the previous component of Need. What ends and objectives will the group pursue collectively? What are the goals of the group members individually?
- Composition - How several members will there be in the group? What are significant points of similarities and differences flanked by them?
- Structure – Specific arrangements to facilitate the conduct of the group, especially in conditions of time and lay.
- Content – What will actually take lay in the group?
- Pre-group get in touch with

If the group membership is predetermined, the procedure starts from composition, for instance, all female students learning in a scrupulous class of a school or an whole floor of a residential institution. The worker then tries to ascertain the needs, capabilities and motivations of the persons mandated to be in the group; determines the purpose, structure, and content and engages in pre-group contacts. Components of Purpose and Composition listed above are explained further.

Purpose

There is need to discover the best fit in the middle of dissimilar perceptions of purpose of the proposed group – the purpose envisaged by the agency, the group worker, the individual member and the group as a whole. Purpose defines the kind of the proposed group:

- Socialization and growth-oriented groups- to develop members' competence in region of general need, to enhance personal growth and to cope with challenges of developmental tasks;
- Support and self-help groups: to give peer support and mutual aid in relieving stress related to hard life situation, to fight discrimination and enhance self-esteem when persons are stigmatized as a result of other persons' lack of understanding or prejudice concerning their ethnicity, situation, illness, or behaviour.
- Task groups, teams and social action groups: to accomplish a scrupulous task. The boundaries of purpose in Growth-oriented and task groups may often get blurred overtime.

Composition

Composition refers to the number and features of both members and the worker who will participate in the group. An understanding of need and a tentative formulation of purpose are key determinants of composition.

A widely accepted principle of composition is that the group should be *similar* (homogenous) enough to ensure commonality of need / interest and compatibility but disparate (heterogeneous) enough to ensure that members will contribute to each other's benefit.

Referring to number of studies, Benson states that a group is more effective if members have *similar descriptive attributes* (age, caste, gender, residence or geographical proximity, marital status, disability or hard social situation, job or socio-economic status, etc.) as this encourages cohesion, inter-activeness and compatibility.

On the other hand, *dissimilar behavioural attributes* (skill to communicate, motivational stages, skill to relate to others, degree of disturbance etc.) will be useful to the group in fostering interest and responsiveness. Knowledge of interests (assumed, expressed or implicit), needs, motivations and aspirations and their social milieu helps worker to create reliable assessment of potential members to expect members' behaviour and responses in the proposed group. In the therapy groups, extensive formal intake interviews are required to identify the members likely to benefit mainly from the proposed group.

DECISIONS TO BE TAKEN

Much before the group is brought into subsistence or shaped and the method of group work applied, the professional social worker has to discover answers to some significant questions and take sure decisions.

- What are the aims, mission and programmes of the social agency or any other auspices within which social worker is operating?
- Who constitute the target groups of the social agency?
- Is group work the best method to help the target group? Can individuals better be helped through use of case work? Are there some needs, troubles or concerns which are shared by a number of people for whom some general aim can be recognized? Is the group likely to be a useful and efficient device for helping the target population? To decide to use groups because it is comparatively economical (vis-à-vis case work) is a mistaken notion. Group work has its own demands on the worker and the members which require serious consideration.
- Is by groups to help people feasible in conditions of required possessions in conditions of time, legroom, equipment, finances or skills, necessary support and social environment? What are the likely costs and benefits of joining the group for the prospective members and for those outside the group?
- Since usually the professional social workers are not trained in the use of specific methods of social work (especially in generic courses), the worker scheduling to form the group needs to ascertain his / her own expertise and comfort stage in working with groups.
- What is the specific purpose of the group under consideration? Who decides the purpose for the group – the agency's service delivery organization, the worker's perception or judgement, some social group or an individual in the society?
- What benefits can the worker hope to procure for the persons drawn from the target population? The purpose of the group may develop from the client group of the social agency or from the agency's decision itself.
- Who are the prospective members of the group? How will they be selected and enlisted in the group?
- Are you acquainted with the population group from where the members will be drawn? Often, while working in a society, a

residential institution, hospital or school, the worker may be well-known to the persons in the target group. Though, the decision to form a group will require deeper insight into potential members' personal features and their capability to benefit from group experiences.

- Will the selection of the members be on the foundation of -
  - Self-selection where individuals are free to approach jointly on their own, such groups tending to be of people recognized to each other, having affinity in some region of general interest.
  - The worker deliberately selects prospective members according to their interest, ability, need, problem or concern keeping also in mind the purpose of the proposed group;
  - Membership is mandated by some power as in the case of delinquents / probationers mandated by the Courts; or the Principal of a school directing students occupied in indiscipline to statement in a group;
- The group worker needs to create detailed scheduling decisions in relation to the structure and character of the group. "Thinking out a appropriate group structure for a population of persons is a fairly intricate task which requires an understanding of the population, information in relation to the range of groups which can be intended, and an appreciation of the demands which dissimilar structures create on group members and of the experiences which dissimilar structures are likely to generate"
  - What is the expected duration of the group in conditions of number of sessions after which the group is expected to disband or terminate, or it will be open-ended as it very often happens in groups in the open society, or in residential organizations where the inmates' population is fluid because the inmates may get discharged and new inmates join in.
  - What will be the duration and frequency of the sessions, for instance, weekly sessions of one or one-and-half hour each.
  - What will be the size of the proposed group? How many members will be enrolled? Social group work is more effective in small groups as it facilitates interactions, communication and efficient role performance. Usually the 'preferred size is 7-8 members but the range of 5-10 members is acceptable'. Though, depending on the purpose, a relatively superior group also may work well. Another determinant of the size is the prospective members' capability to interact in groups. For members with limited interactional skills, the size of the group could be limited to 4-5 persons

— Decision in relation to the composition of the group has to be made right at the scheduling stage. We have already referred to the issue of homogeneity and heterogeneity in the middle of the members (*i.e.*, parallel or dissimilarity beside social categories like age, gender, minority status, ethnicity, social troubles, etc) as also the factors which may power members' willingness to join in the proposed group. In therapy oriented groups size and composition are of vital significance.

— What 'programme' will be followed in the group – discussions – open or topic oriented; behaviours – games, art and craft work, drama, role plays, simulation, society service etc.; get-togethers or combination of these?

— What possessions in conditions of legroom, time, finances and technology will be required by the group? Are the similar accessible in the agency / society or do they need to be generated?

- Does the worker need to make pre-group contacts with prospective members or engage in some preparatory work with them?
- Deciding how to open the group, how to monitor the group through its life span, how to evaluate the group's performance and development, and when and how to terminate the group is equally significant for efficient scheduling of a group.
- In several of the non-social work organizations like schools, hospitals, prisons the social worker needs co-operation and support of staff belonging to other disciplines. It is significant to ensure that the proposed group will receive necessary support from other staff of the agency.
- The worker has to ascertain whether s/he needs permission of some power for launching the proposed group. In case of society based self-help groups, support and sanction of society leaders is often vital for the success of the group.

PROCEDURE OF GROUP FORMATION – TASKS UNDERTAKEN BY THE SOCIAL GROUP WORKER

We mentioned in the forgoing discussion that the worker has to discover answers to sure significant questions while embarking on the use of a group to help the target group. Now we will outline the actual tasks the worker has to engage in to initiate the procedure of group formation.

- Once the use of group has been accepted as the best possible option (after considering dissimilar alternatives) for helping the designated population group, the group worker has to formulate a tentative purpose for the group. The worker identifies a general need or concern of the target group and translates that need into a tentative purpose of the prospective group. The following situations may illustrate the rationale for selecting group work as a better option to help:

— Many people facing similar situation can benefit from sharing their experiences (parents of mentally challenged children);

— Persons belonging to similar stage of development like adolescents who can benefit from positive group experiences;

— When individuals have troubles with power figures, in relating to others or having problem of separation;

— When the target for change is in the social environment like sanitation, educational or health services and the need is for enhanced civic amenities;

— When people wish or need to avail of the benefit of some beneficiary-oriented scheme of the govt. which requires formation of a group, like self-help groups for procuring micro credit.

- If the worker is required to work with an already existing group, s/he needs to comprehend its purpose before s/he begins work with the group.
- Achieving optimally effective composition of the group is a crucial task throughout the formation procedure. Will the group constitute only of women or young girls, only of men or male youth, of children, of able-bodied or differently-abled persons; will it be a mix of persons belonging to dissimilar regions, religions, words and cultural background, with dissimilar educational and socio-economic status? This decision will mainly be based on the tentative purpose formulated, anticipated member behaviour and emotional response to prospective members and the knowledge of the patterns of habitual social intercourse prevailing in the superior population group. For instance in Indian society, groups with the mix of genders are socially not very acceptable, unless the group's purpose is task-oriented (for instance in the corporate sector) which may benefit by a combination of diverse skills and perspectives; or a group composed of all aggressive or all withdrawn persons may not benefit from interaction with each other. Who is selected to be the member of the proposed group has lasting impact on the individual member and the group as a whole. If persons are placed in groups that are unsuitable for them, they may be harmed by the membership or may drop out of the group. Dissimilar results flow from different combination of people. What is significant is to see that there is a good fit flanked by any one person

and the other group members. The size of the group should be determined by the nature of interaction in the middle of the members and their participation necessary for achieving the purpose of the individual members and the group as a whole.

- The worker identifies potential members from the designated population group on the foundation of dissimilar criteria mentioned earlier.
  - The worker may select members from students of a class; from a list prepared on the foundation of a survey of the families below poverty row or of children afflicted with some specified disability; from the census statement; from membership of previous groups; from the list of persons who may have applied for a scrupulous course or training; or the official records of courts, hospitals, therapists, psychiatric clinics etc.
  - The worker may need to advertise (by word of mouth, leaflets, posters, ads in the newspapers, letters to the organizations like schools from where the potential members can be drawn), state the purpose of initiating the group and invite members to join.
    - In a society-based agency, the worker talked to the society leaders, local MLA and Municipal Councilor, visited the potential members personally and explained the purpose of the group and its likely benefits for them.
    - In an epilepsy clinic for children, parents accompanying their wards were contacted by the social worker and the purpose, content and structure of the proposed group was explained to them personally.
    - In a residential 'Home' for destitute children, the worker, though recognized to the inmates on explanation of her dissimilar responsibilities, explored the strengths and liabilities of the prospective members in conditions of the purpose envisaged, the contribution expected of the members for the proposed group and informed all potential members in relation to the group to be launched and its purpose and structure.
- Pre-group contacts form a important part of the worker's tasks throughout the group formation procedure. The aim of these contacts is to secure appropriate members for the group being planned and their preparation for participation in that group. Through these contacts, the people in the target population get to know the availability and nature of the group based service, their eligibility for that service, ascertain if their goals are similar to others to be met

through the group and prepare them for entry in the scrupulous group. The abovementioned tasks may require more than one visit or meeting. The worker informs the potential members the tentatively planned group structure, the duration and the frequency of the sessions. In case of a group of children or young adults, pre-group contacts also entail meeting the parents / family members to apprise them of the proposed group, its purpose, the benefits it will offer and its structure. Pre-group contacts may happen throughout a meeting of the society residents and society leaders or throughout some social / religious event throughout which the worker gets an opportunity to introduce the group to be launched.

- Pre-group interview (one or more than one) with potential member helps clarify and alleviate member's anxiety in relation to the group's structure, expectations of the group in conditions of his role, response of other members towards his membership. It brings out valuable data in relation to the member's attitudes, perception of his need, life situation or difficulty; and his capability to relate to and to communicate with others; In therapy groups, Intake interviews are held with individuals to arrive at an in-depth assessment of the nature and severity of the problem, difficulty or situation; on the foundation of which their compatibility or incompatibility to group membership may be decided.
- Even when the potential members are influenced of the purpose of the group, they may be anxious to know as to what exactly is expected of them, how the group work and what will the other members will be like.
- Pre-group get in touch with also initiates worker – member connection, whether mannered through an individual interview, a visit to the locality of the potential members, or in a meeting held in the society.
- Sure rules and norms that the members are expected to observe initially, the issues of confidentiality, democratic participation, antidiscrimination, and the manner of modifying the norms need to be shared with the potential members throughout pre-group contacts.

Right from the point of worker / agency agreeing to plan the group; spelling out its purpose, structure, membership; undertaking publicity; selection and recruitment of potential members, worker has to perform numerous significant tasks which have extensive lasting impact on the group's success in conditions of attainment of its goal – individual member's and the group. Scheduling the whole procedure of group formation before the group has its first meeting is vital for the group's success.

PRACTICE PRINCIPLES: GUIDELINES FOR GROUP FORMATION

A number of practitioners and authors have formulated Practice Principles for effective and efficient group work practice. Some of these are particularly concerned with the group formation stage of group work. These principles offer guidelines to the worker for securing effective formation of the proposed group.

- Comfort stage of the worker - Since the professional social workers are usually not trained in the use of specific methods of social work (especially in generic courses), the worker scheduling to form the group needs to ascertain his / her own expertise and comfort stage in working with groups. Whether the worker needs to enhance his / her knowledge and skills before embarking on group formation or to enlist the help of a more experienced co- worker is an significant issue to be resolved.
- Scheduling for a group's formation is a mainly vital step in the group based service to the target population. To avoid its disagreement with the 'value' of client's right to self-determination the plan should not be rigid. In information, flexibility and creativity are two essential attributes of scheduling undertaken for group formation.
- Secure organizational support and sanction for group, and address organizational resistance to groups if needed.
- The purpose of forming the group in view should be clearly for the benefit and service of the target population and not merely to receive aid and grants from some financing organization, to collect data for an undisclosed purpose, or to project sure image in the society. This type of hidden agenda involves manipulation of the target groups and is unethical practice.
- Develop and articulate verbally and/or in writing a clear statement of group purpose that reflects member needs and, where appropriate, agency mission. (AASGW) According to Newstetter, "the deliberate formation of a new group should be encouraged only after the careful review of three thoughts: a) the interests, abilities, experiences and needs of each prospective individual member, b) the agency's purpose, c) the availability of appropriate facilities, leadership and supervision".
- The workers often feel uncomfortable in honestly admitting of the purpose envisaged by them or by the agency. They feel that the purpose as it appears in the blue print of the plan may not sound too attractive to the prospective members and, so, speak of the purpose in conditions which they think may be acceptable to the prospective members. This is likely to make mistrust, confusion and anxiety in the middle of the potential members.

- Particularly in the case of involuntary groups (mandated membership like in groups under court order or students under suspension for indiscipline), the clarity in the statement of purpose, the proposed group structure and content is very vital to win prospective members', participation, interaction and trust.
- Develop and articulate clear statement of worker role that reflects the group's purpose.
- Use preparatory empathy to tune into members' feelings and reactions to group's beginning.
- Tuning-in helps the worker anticipate members' needs and feelings and develop preliminary empathy. The worker also tunes-in to one's own feelings, attitudes and thoughts in relation to the ensuing group encounter.
- Set up meeting lay, time, etc. that promotes member comfort and cohesion.
- Negotiation and contracting with the members in the context of purpose, structure and norms of the proposed group contribute significantly to future stability and cohesiveness of the group.
- Groups need to be shaped in a method that enables them to do the work that they will be asked to do. This involves minimizing barriers to group cohesiveness and then giving them the possessions they need (in conditions of talent, time, etc.).
- For the formation to be efficient, group workers need to take into explanation any constraints that can power the performance of the group as a whole and that of the individuals within the group, such as their previous experience, gender, ethnicity, and interests.
- The worker needs to pay special attention to composition issues to achieve a balanced group.
- The worker is accountable for his /her professional expertise which is made accessible for the service of the people. While formulating Standards for Social Work Practice with Groups, AASGW recommended the following regions of knowledge for the group worker at the pre-group stage:
  - Organization's mission and function and how this powers nature of group work service
  - Social and institutional barriers which may impact on the development of group work service
  - Issues associated with group composition
  - Human life cycle and its connection to potential members' needs
  - Cultural factors and their power on potential members' lives and their skill to engage in group and relate to others

- Kinds of groups and their connection to member needs
- Specific kinds of individual and social troubles that lead to a need for group

CONCLUSION

Getting a group shaped takes time and skills. Described variously as the group formation stage, pre-group or pre-affiliation stage, it requires thoughtful and serious scheduling and its rigorous and patient execution before the first meeting of the group is ever mannered. Group formation constitutes of a series of interconnected behaviours , which are based on a judicious and well thought out plan. Society, social agency, the social group worker and the people availing of the group-based services jointly determine what type of group is planned and launched; who join the proposed group and why; and what is going to happen in the group in the forthcoming era. Prior to the first meeting, a conception of the necessity and utility of group experience has to be framed in the contexts of the worker / agency perceptions, which in turn, is based on the worker's personal and professional experiences, knowledge of human development and social systems. Other relevant factors follow from this initial conceptualization. The major components of the group formation plan embedded in the social and agency contexts are need, purpose, and structure, content and pre-group contacts.

Worker has to engage in a number of tasks to recruit and prepare potential members for the proposed group. How balanced a group membership is supervised by the worker will ensure, to a great extent, the success of the group in conditions of attainment of individual / group / agency goals. You have learnt practice principles particularly relevant to pre-group or group formation. These principles go a extensive method in guiding group formation procedure beside effective and ethical pathways.

8

Yarn Products' Production

YARN DESIGN

YARN AND YARN DESIGN FROM FIBRE TO PRODUCT PURCHASE

The Global Yarn and Clothing Industries

Yarn making is a very ancient craft, with a history almost as old as mankind itself. Remembered and recorded in poetry and ancient stories and myths, yarn have always been important to man. As well as providing protection from the elements, the first yarn were used as decoration, providing status for the owner. They were also used as tools; bags for transporting belongings and for holding food as it was gathered. Yarn are produced in almost every country of the world, sometimes for consumption exclusively in the country of manufacture, sometimes mainly for export. From cottage industry to multi-national corporation, yarn and clothing are truly global industries. In 1782, the invention of the steam engine gave the world a new power source and started the Industrial Revolution.

Previous to this the production of yarn had been a domestic system, a cottage industry with yarn spun, knitted and woven in the home. By the middle of the nineteenth century, however, there was a whole range of new machines and inventions that were to take yarn into an era of mass production in factories. The development of man-made fibres and new dyestuffs in the early part of the twentieth century, and continuing technological developments, have led and continue to lead to new products and applications.

The actual processes of yarn manufacture, however, are still very much as they have always been, with the vast majority of cloth being woven or knitted from yarn spun from fibre. And, while much production may be very technologically advanced, hand-produced yarn are still made in many countries exactly as they were many, many years ago. Nowadays, many different types of companies are involved in the production of yarn and clothing world-wide;

some companies own many huge manufacturing plants in many different countries while others will have only a few employees and some may not actually manufacture at all.

Yarn Materials, Processes, and Products

Fibres are manufactured or processed into yarn, and yarn are made into fabrics. Fabrics may be manufactured by a variety of processes including knitting, weaving, lace making, felt-making, knotting, and stitch bonding. These fabrics may be industrial yarn with detailed technical and performance specifications, or they may be sold either to retail or contract as apparel, furnishings or household yarn, where aesthetics may be as, or sometimes even more important than performance.

The fabrics may be coloured by dyeing or printing, or be finished to enhance their appearance or performance. A wide diversity of products are made from yarn products or have some yarn components; yarn go into car tyres, and geoyarn are used for lining reservoirs, while medical applications include artificial ligaments and replacement arteries.

Design in yarn and clothing

Every yarn product is designed: that is, it is made specifically to some kind of plan. Design decisions are made at every stage in the manufacturing process—what fibres should be used in a yarn, what yarn in a fabric, what weight of fabric should be produced, what colours should the yarn or fabric be produced in, what fabric structures should be used and what finishes applied. These decisions may be made by engineers and technologists in the case of industrial or medical yarn where performance requirements are paramount, or, more often in the case of apparel, furnishings and household yarn, by designers trained in aesthetics, technology and marketing.

The designers found in the yarn and clothing industries are frequently involved throughout the design process, from initial identification of a need/ requirement, through research, generation of initial design ideas, design development and testing to ultimate product specification.

Designers found in the yarn and clothing industries

The designers found in yarn and clothing include:

- Colourists predicting and forecasting future colour ranges
- Yarn designers
- Knitted fabric designers
- Woven fabric designers
- Carpet designers
- Print designers

- Embroidery designers
- Knitwear designers
- Garment designers
- Accessory designers
- Print producers
- Stylists
- Colourists developing colourways
- Repeat artists

Fibres

Fabric is made from yarn, and yarn is made from fibres. These fibres can be either natural or man-made. Natural fibres include animal fibres (*e.g.* wool and silk), vegetable fibres (*e.g.* jute and cotton) and mineral fibres (*e.g.* asbestos). Man-made fibres are either regenerated or synthetic; viscose rayon, based on regenerated cellulose, is man-made but not synthetic while polyester, polypropylene and nylon are all synthetic fibres. Synthetic fibres are produced by the large chemical companies including Dupont, Bayer, Hoechst and Astra Zeneca. Many of these companies produce no fabric but specialise in the production of certain types of fibre which they sell on as fibres or manufacture into yarn.

Yarn

Yarn producers or spinners buy in natural and/or man-made fibres to make these into yarn of different sizes and characters; regular and fancy yarn. For many years the main spinning systems could be given as woollen, worsted and cotton, and these systems gave rise to the woollen, worsted and cotton industries.

Developments in spinning, however, have led to new spinning systems including 'open-end', 'self-twist' and 'jet' spinning. At its simplest, yarn production is essentially about taking fibres, organising them so that they lie in a lengthways direction and twisting them to create a yarn. By combining fibre types, and using different spinning systems and machinery, yarn can be developed with individual profiles suitable for a vast range of end uses. Regular yarn are those which have a regular straight profile and these can be twisted together, making 'two-fold' or 'three-fold' yarn for example. Fancy yarn can be created by deliberately introducing irregularities or intermittent effects along their length.

Yarn can be combined together as components of new yarn with different effects and properties from their component parts. As well as changing the appearance of a fabric, the introduction of a fancy yarn will affect the handle and performance of that fabric.

Woven fabrics

Strictly speaking, the definition of a yarn is 'a woven fabric' but the term yarn is now considered to cover any product that uses yarn materials or is made by yarn processes. Essentially, woven fabrics are structures produced by interlacing two sets of threads; the warp which runs in a lengthways direction and the weft which runs in a widthways direction. Weaving methods include tapestry and jacquard.

Knitted Fabrics

Knitted fabrics are produced by interlacing loops of yarn. In weft knitting, loops are formed one at a time in a weft-ways direction as the fabric is formed. Hand-knitting with a pair of knitting needles is weft knitting. In warp knitting there is a set of warp yarn which are simultaneously formed into loops. To connect these chains of loops the warp threads are moved sideways in such a way as to cause the loops to interlink.

Lace and non-woven Fabrics

Fabrics may also be produced by methods other than weaving and knitting. Lace is an open-work fabric made by looping, plaiting or twisting threads by means of a needle or a set of bobbins. Fabrics produced by crochet and macramé are often called lace, although strictly speaking they are not. Knotting is another way of making fabrics.

Knotting was a popular pastime for women in eighteenth-century Europe and colonial North America, and one method still seen today is macramé. A knotting process is also used for fishing nets, and some rugs and carpets are knotted — made by tying yarn onto a foundation weave. There is also a group of fabrics called non-wovens which include true felt (where animal fibres are matted together) and fabrics produced by bonding webs of fibres together by stitching or by sticking with adhesive. However, in terms of volume produced, knitted and woven fabrics are by far the most common methods of fabric production.

Fabric Terms

A length of woven or knitted fabric is usually referred to as a 'piece'. Often, fabric woven by a mill will not be coloured and this undyed fabric is called 'grey cloth'. Colour can be added by dyeing the piece, and such fabric is referred to as being 'piecedyed'.

Colour can also be added to a fabric by applying pigments or dyes in a printing or other colouring process after weaving or knitting, or by using already dyed yarn in the construction of the fabric. Cloth made from dyed yarn will not normally be dyed again or printed. 'Finishing' is what happens after the fabric has been made. The finishing processes employed will be determined by the type of fabric and its performance requirements. Any excess dye will normally be removed, any applied pigment will normally be set, and any dye will be fixed. Fabrics may be brushed or raised to enhance appearance and handle, or fire-retardant and soil-resist treatments may be applied. Fire retardancy may be a product performance prerequisite; anti-soiling and anti-static finishes, while not necessarily pre-requisites, enhance performance, as do methods of coating fabrics to produce microporous surfaces.

Geography and Fabric Types

Certain countries, and areas within countries, have developed industries around specific fibres/fabric types and there are still parts of the world where craftsmen produce fabrics exclusive to them, such as some hand-crafted batiks and weaves. In the UK, Manchester was nicknamed 'Cottonopolis' as it was built in the main (as was much of Lancashire) from money earned through the cotton trade.

Scotland developed a woollen trade through both woven and knitted fabrics, while, again in the UK, Yorkshire was home to the worsted industry and the Midlands became famous for knitting and lace making. India has a history of cotton manufacture and, in the eighteenth century, was famous throughout Europe for its mordanted cottons or chintzes.

YARN ORGANIZATIONS

Size and Structure

The factories producing yarn are usually called mills. Some mills take in fibre, spin yarn, dye these and then either weave or knit these into fabrics/garments. Such organisations that are involved in several yarn processes are described as 'vertical'.

There are, however, also plenty of companies (often smaller but not always) specialising in one of these functions, usually on a commission basis, and these organisations are described as 'horizontal'. There are many examples of commission knitters, printers, dyers and finishers — companies that for a unit price process yarn goods for other companies in the industry. Many yarn organisations today are huge multi-national corporations involved in more than one yarn process; producing fibres, spinning, dyeing, weaving and knitting, printing, and garment manufacture.

These companies will often have these various processes carried out in many different countries. It is therefore not unusual to buy a garment in Japan or the USA that was made up in Portugal, with sewing threads from the UK, from fabric woven in Korea, from yarn manufactured in Italy and from fibres made in Germany. Until the 1960s and 70s, most yarn companies in the UK, Europe and the USA were relatively small organisations. Many were vertical operations, involved in all the manufacturing processes from fibre to finished product, although there were some horizontal organisations, specialising in only one process such as spinning, weaving, or dyeing and printing. There was a period of major change in the 1970s when many mergers and take-overs took place resulting in re-groupings of operations. Yarn and clothing in the twenty-first century will continue to be a truly global industry.

Converters and Wholesalers

Manufacturers of grey cloth may sell this fabric to converters rather than do anything further to it themselves. Converters buy grey cloth and convert this by having it dyed or printed, and then finished. A mill will own specific equipment, or plant, which must be kept operating to maintain profitability; a converter has greater flexibility in that such an operation does not need to own any equipment, having everything done by other organisations. If a converting company has a new idea, they can find a new resource without compromising existing business.

A mill, however, with all its operations under one roof, does have more control. As we have seen, fabrics are usually produced by the piece. Mills and converters usually sell by the piece. Wholesalers essentially buy from a manufacturer and, without changing the product, sell it in smaller quantities to retailers or smaller manufacturers.

CATEGORISING YARN

Yarn can be categorised by production company, by end-use, and by the market for which they are designed.

Production Companies

The companies involved in yarn and clothing production can be grouped in different ways, for example by product or by manufacturing process. Many companies fall into several categories. A mill may produce fabrics but it may also convert fabric that it is uneconomical for it to produce itself. Usually labelled by the function which they primarily perform, *i.e.* a spinner, a knitter, a weaver, etc., companies may also be known for the end-use of their fabric, *i.e.* as a producer of men's knitwear, lingerie, contract furnishings, etc.

Mills often produce fabrics for many different end-uses while converters mostly develop fabrics exclusively for one end-use. A company might describe itself as: 'Hosiery manufacturers, and spinners and doublers of super merinos in white and colour. Also manufacturers of high quality underwear and knitwear for men and women.'

End-uses

Yarn are found in a hugely diverse range of products. Clothing us from birth until death, yarn protect us and make us feel good. Our homes are made more comfortable by yarn that keep in heat and by yarn that shield us from the sun. Keeping us warm at night, yarn also dry us when we are wet and can support injured limbs. Yarn allow us to make tea directly in a cup. More recently specialised yarn have been developed for medical use as artificial replacement ligaments and arteries, and geoyarn are used in the construction of dams and motorways and even bunkers on golf courses.

Apparel Yarn

The clothing or apparel market includes most garments that are worn. A huge consumer of fabric, clothing manufacture can be split by market, *e.g.* men's, women's and children's clothing, sportswear, casual wear or formal wear. However, not all fabrics for garments are considered part of the apparel market. Fabrics such as the specialised protective clothing for fire-fighters, pilots and those in similar hazardous occupations are considered part of the industrial yarn market, and specialist clothing for leisure and ski wear, etc. are considered as being consumer yarn.

Furnishing Fabrics or Interior Yarn

The furnishing market is another huge consumer of yarn, for curtains, upholstery fabrics, carpets and wall coverings, either domestic or contract. Domestic furnishings are those found in the home, while contract furnishings are those used in offices and public buildings such as schools, hotels and hospitals.

Household Yarn

This category includes all yarn products used within the home except furnishings, including sheets, pillowcases, towels, blankets, tablecloths, etc. When these products are used in the contract market they may be referred to as 'institutional fabrics'.

Industrial Yarn

Car tyres, medical yarn and geoyarn are all examples of industrial yarn. Industrial yarn also covers such yarn products as filters, conveyor belts, car safety belts and parachute cords. Performance is of prime importance in this category.

Consumer Yarn

This category could be described as including any yarn not falling into the previous categories. Recreational items such as tents and back packs may be referred to as consumer yarn, as well as awnings and umbrellas and luggage. Although in this category performance can be very important, aesthetics can be equally so.

Yarn Categorised by Market Area and Price

Yarn and yarn products can be categorised by the market area for which these are intended and by price. Expensive fabrics, apparel and furnishing products may be described as 'upper end', 'top', 'exclusive', 'haute couture' and 'designer'. However, the largest quantity of fabrics and yarn products are sold in the middle volume, or mass market area, and in the lower, down-market area.

Categorising Yarn Companies

Any yarn company can be described using various labels — by the manufacturing process carried out, by the product type and by product market area and price. Companies may also be known as volume converters, top-end fashion-fabric producers, cut and sewn or fully-fashioned knitters, etc.

YARN DESIGNERS

The Diversity of Yarn Design and Yarn Designers

The diversity of the yarn and clothing industries is reflected in the many different types of designers needed. At every stage of manufacture of yarn there are colourists determining the fashion colours in which the fibres will be produced, yarn designers developing yarn to meet certain requirements, knitted-fabric designers, woven-fabric designers, carpet designers, print designers, embroidery designers, knitwear designers, designers of women's wear, men's wear and children's wear, accessory designers, designers of casualwear, sportswear, eveningwear, swimwear and designers for the mass market, haute couture and designer labels, etc.

The Purpose of the Yarn Designer

The role of the designer can be quite complex but the overall purpose can be stated quite simply —the yarn designer has to design and produce, to an agreed timetable, an agreed number of commercially viable fabric designs. Depending on the markets that he or she is designing for, several different activities are involved in fabric design and the number and type in which any designer is involved will vary according to the product and production methods used, and the type of company for which the work is done.

Stylists

Designers also put together ranges. For example a stylist might handle the development of a company's range of printed fabrics. A range is a group of fabrics (or products) designed, developed and edited to be shown and sold to the market each season. The stylist initiates the design work, organises and directs the development and colouration of intended designs (frequently using freelance designers), and co-ordinates with manufacturing personnel to have samples made.

These samples are shown to customers; the stylist then edits and finalises the group of designs that will form that season's range. Further down the chain, buyers and merchandisers in retail organisations do much the same range-building processes.

Colourists

Some designers work purely with colour, predicting colour trends and putting together palettes of colours for specific seasons and product groups. Other colourists will work further down the design process line, colouring designs produced by other designers to create different and alternative colourways.

Repeat artists

Acompany producing printed yarn will often employ designers whose main function is to take designs and put these into a size and repeat appropriate to the intended enduse.

In-house and Freelance Designers

Designers may work for manufacturing companies as in-house designers, or they may work independently as freelance designers. In-house designers, or as they are sometimes called staff designers, are employed by a company usually on a full-time basis, although some may be employed part-time. Often they work within a manufacturing environment, although they can also be employed by retailers and by converters.

Freelance designers may either work for independent studios or through an agent, producing designs on paper for which the studio/agent receives a commission when the designs are sold to mills and converters. Alternatively, freelance designers may put together a portfolio of their designs, which they may sell directly to stylists.

The work they produce for their portfolio, while it will have at least to reflect trends, will be often very much what they themselves like and want to produce. While the designer may have a view of the type of customer who will buy their designs when the design is developed, there may be no specific customer waiting to buy their work on completion. Freelance designers may also develop design work according to a stylist's specification. For example, a freelance print designer who is particularly good at intricate florals may well be approached by a stylist to work on a specific print idea that will form part of that company's new season's collection. The brief may include size details, colouring details and even the type of flowers to be painted. This work will be commissioned in advance.

The designer develops their paperwork with the knowledge that when it is finished there is a buyer for it. Freelance weave and knit designers will normally work on a specific project with a manufacturer. They will be commissioned to produce a range of fabrics, or, in the case of a knitwear designer, a range of knitted-garment designs. A third type of designer found within yarn is the consultant designer. A consultant is employed by a company to advise on design matters and may be given the task of managing the design programme.

A consultant designer will usually work for several companies at any one time, although their contract may be such as to impose restrictions on their working for closely related organisations.

Very often consultants will do little actual working-through of design ideas themselves; rather they will make design policy decisions and direct other designers who may be in-house or freelance. All designer systems have advantages and disadvantages to the designers themselves and to the organisations for whom they work.

Just as many different people doing different jobs are given the name designer, so too are many rooms called design studios. The 'studio' can be anything from an area set aside on the factory floor to a large, smart office or even suite of offices.

One designer or a whole team of designers can work in a studio and a few or many of the functions of, and stages in, the design process might be carried out there.

Timing in the yarn and Clothing Industries

Products are planned and produced well ahead of retail selling seasons in all areas of the yarn and clothing industries. In August, and even July, autumn clothing appears in the shops. Before this clothing can be designed, fabrics must be designed and shown to clothing manufacturers, and orders placed, produced and delivered.

Before then, new yarn and sometimes even new fibres must be developed. The scheduling and amount of time necessary for all of these steps are dependent on the amount of change occurring in the product, the volume being produced and the efficiency of the producing companies. At a minimum, yarn are usually designed a year to a year-and-a-half in front of the retail season. Major changes such as the development of a completely new type of fabric will take even longer.

Seasonal Ranges

Manufacturers producing for the fashion industry will normally launch two main ranges a year; for the selling seasons Spring/Summer and Autumn/Winter. Sometimes, midseason ranges are also produced. Manufacturers of furnishing fabrics and household yarn normally produce one main range per year. In every segment of the market, the higher-volume manufacturers work further ahead of the retail season than the companies producing lower volumes (at the higher end of the market).

Generally, smaller operations do not require the long lead-times that are necessary for large-volume production runs.

Printed and Constructed Yarn

Yarn designers may be categorised by the types of product or fabrics for which they design. Printed yarn are often considered to include fabrics patterned by dyeing techniques as well as those where the design is applied to the fabric by a printing process. Constructed yarn include woven yarn, knitted yarn, lace and carpets.

From Sketchbook to Fabric Samples

The design process is about the realisation of ideas —ideas that are transformed into tangible products. The designer will start with some sort of brief describing the project. For the in-house designer, the briefing may be very informal as their experience within the company will usually mean that they will already have an understanding of the requirements of the range. They may, indeed, be the person setting the brief! Hopefully, any brief, formally or informally delivered, will have all the information needed, but if not, the designer must be ready to ask questions.

Who is the customer? What is the price range? What are the manufacturing possibilities in terms of machine availability? The designer needs to get as much information as possible. Designs come from a variety of different sources, and different designers will have different methods for developing design work from initial ideas.

Most designers keep some form of sketchbook —something they carry with them and use to record and store ideas to call upon as a source of inspiration in the future.

This may contain drawings and photographs, magazine cuttings and fabric pieces. Often, a brainstorming session will be used to give inspiration and generate a wide variety of ideas. Some of the considered themes and ideas will be selected and subsequent artwork based around these. While the way these ideas are used will be individual to the designer, much of the process is similar for many different types of designers.

Yarn Design

Yarn designers have to know what machinery they have available to them and what that machinery is capable of producing. They will usually start with ideas collected on their travels; colour ideas, texture ideas, looks and moods that they see as becoming important in the future. They should have an understanding of their existing and potential customers and what they want. The development of a range of yarn may involve the creation of a completely new yarn using a newly-developed fibre.

It may involve the tweaking and fine-tuning of existing production ranges to meet the requirements of a future market. An existing yarn quality may require to be re-coloured so that it can be used to service a fashion market. Some ranges of yarn will run from season to season servicing a classic market area. For example, a range of knitwear yarn for a manufacturer of schoolwear will include the classic school colours — navy, maroon, grey, bottle green, bright red and bright blue.

The knowledge base that the yarn designer will have built-up is the springboard from which they work, in conjunction with technicians and development personnel, looking at the overall range, determining qualities and the types of yarn appropriate to their market area. The selling cost of the yarn must also be considered.

Colour is very important and yarn designers will often subscribe to colour prediction publications to help them determine their colour palettes. The same colours or shades of colours will often run through the different yarn types. Co-ordination is as important in yarn design as in any other area, helping the customer to see how the product can be used and often prompting more sales. Yarn may be designed to work with other qualities in the range.

Sample yarn will be specified and made using different components and in different colourways until the designer is happy with the handle and appearance. The yarn may have to perform in specific situations and they will have to be tested to ensure that any such performance requirements are met.

Weave Design

After the briefing meeting where the work that is required is established, and after initial ideas have been collected, one of the first things that a weave designer will do is to establish a colour palette. This colour palette may change as design work develops but such changes are normally minimal. A colour palette will usually consist of groups of colours, chosen with regard to trends and predicted directions.

The way these colours work together will be important. Adegree of flexibility is usually desirable as economic constraints will normally demand that stocks of yarn (both for sampling and production) are kept to a minimum.

Once an initial colour palette has been decided upon, suitable yarn are then selected.

These will normally be selected from ranges offered by yarn producers and frequently from the colours in which the yarn producers have decided to run their ranges; sometimes, however, colours will have to be specially dyed. From their initial paperwork, weave designers will usually take their design ideas and develop them on the loom. Sample warps will be made up and different weave, colour and yarn combinations tried. Shape and pattern are important but these might well evolve more with the weaves used. The way woven designs repeat is determined by the way the yarn are threaded up and the way these threads are lifted.

This is a very integral part of the weave design process, and a good understanding of fabric structures is vital to allow the weave designer to be as creative as possible within the constraints set by the type of loom that is being used and its patterning capabilities. Sample blankets, or 'section' blankets as they are also known, consist of different sections across a warp. These sections may vary in colour, yarn type and/or weave.

Initial sample warps will often have several warp sections with different drafts, and different warp patterns and yarn. Different lifting plans (which, along with the way the loom is threaded, determine the weave) will be tried out as will a variety of weft patterns and yarn, again in sections. As well as true designs where the planned warp will be woven with the intended weft, there will be many areas where crossings occur.

These are sections where warp and weft patterns combine more by chance than by design. Areas of interest will be developed and hopefully an initial sample blanket will provide plenty of ideas for subsequent development. Having considered initial ideas, further sample blankets will be developed; these will show full repeats of the proposed fabric designs and will often have sections in alternative colourways.

Some weave designers will work on computer-aided design (CAD) systems, using these to try out ideas before the weaving stage. Many of the sophisticated weave CAD systems currently available, as well as showing representations of what the fabric will look like, also have facilities to show proposed weave designs in situ, on garments and on other products.

However, in practice it is still usual to weave actual fabric samples of those ideas which best meet the brief. The way fabric feels, its handle, is still a very important factor in fabric selection. With a selection of ideas chosen, the making particulars or fabric specifications will be recorded so that these designs may be produced again. The yarn and colours used, the draft (the way the loom is threaded), weave, lifting plan (the way the warp threads are raised), warping and picking plans (colour and yarn arrangements in the warp and weft directions) all need to be recorded, as do any subsequent process such as applying a fire-retardant finish or raising. Throughout the weave design process, the designer is constantly having to make decisions, on colour, yarn, weave, size of repeat, etc.

These decisions are made from a knowledge base that includes intended selling price, customer preferences, the range as a whole, co-ordination requirements, performance requirements, and so on. Once the design ideas are developed through into fabrics, these will usually then be presented in-house to the sales team, and then to customers.

Knit Fabric and Knitwear Design

Again, as in weave design, the design of a new season's range will usually start with the development of a colour palette. This will come from initial ideas generated in response to information from a briefing meeting. Suitable yarn will then be selected. The working through of design ideas is very much restricted by the type of knitting machine being used and the type of structures that can be created on it. Inspiration from drawn work will influence the textures, colours and proportions of a knitted fabric. Creating visual representations of sketchbook work is only possible when machinery with jacquard patterning capabilities is used.

Some designers will put their initial design ideas straight into graph form, ready for translation by a knitting mechanic. Others may do more paperwork, perhaps developing fabric and garment styling ideas side by side. The fabric ideas worked out, the designer will then usually decide which yarn should be tried in which fabrics. It is usual to knit several swatches to enable the designer to see how any fabric looks in different yarn and colour combinations. For the knitwear designer, concerned with the design of final garment as well as the fabrics, styling ideas must then be worked out. The designer of cut and sewn knitwear will probably sketch out styling ideas, and the next stage will be to cut a pattern.

A first sample will be cut out and made up, pressed, measured and tried for fit. Any necessary alterations will be made and a second sample made to check that these alterations and the final specifications are correct. To help knitted fabric and knitwear designers, there are specific CAD systems available which allow representations of fabrics to be viewed before knitting. They also often allow fabric ideas to be mapped onto garments, and these facilities should cut down on actual sampling time. In practice, this is not always the case as the ease of changing things on the computer often results in requests for more initial work giving more choice. Once the garments and colour ways that will form the range are decided upon, the range is then ready for presentation to sales teams and customers.

Print design

The print design process can take many different forms. With modern technology, sketchbook work can be taken and re-created on fabric very much as it appeared originally. Print designers, after initial research, will usually work their ideas through on paper. Any medium can be used in the artwork for print design, but materials with which designers can achieve facsimiles of particular desired effects in manufactured fabrics allow design problems to be more accurately resolved in the studio before the design is put into production. Print designs are sometimes worked in repeat from the start but often they are designed in balance and put in repeat later. A balanced design is called a 'croquis' (a French word meaning *sketch*).

A croquis should give the impression that would be seen if a frame were placed over any section of the finished cloth. Although not in repeat, a croquis will give the feeling of being in repeat.

The most common repeat structures are simple block repeats, half drops and bricks. Print designs for apparel fabrics are usually designed in croquis form with the repeat sizes varying enormously. Upholstery and curtain fabrics have standard repeat sizes so these are often designed as a repeat pattern with the size being carefully considered from the beginning. Often, the involvement of a print designer will end with the production of the design on paper, the preparation of colour separations and the approval of these usually being undertaken by the printer.

However, in-house print designers will usually be responsible for approving the strike-offs (sample prints on fabric). Again many print designers (particularly those working for large multinational companies and corporations) will often find themselves working with CAD systems. Such print systems allow designers to develop and expand design ideas readily. Design ideas can be scanned in, and repeat, cut-and-paste and other functions allow ideas to be manipulated quickly.

Many systems also have colour separation facilities to aid in the preparation of screens. Some systems even allow sample fabrics to be produced, and this is a very valuable tool, making it much easier for print designers to take their ideas right through to fabric.

Design Adaptation and Modification

As well as producing ranges, yarn designers will have to adapt and modify existing designs, design ideas and sketches submitted by customers. Here, yarn designers are using their creative and technical skills to translate ideas into reality. They must, of course, be careful not to copy or adapt designs that have copyright, or are registered, without the express permission of the copyright holder or design originator.

Examples of Yarn Design Briefs/problems

An in-house designer working for a manufacturer of printed fabrics may be asked by a customer selling babywear to develop a small design with chickens as the motif. The babywear company may also employ a designer and they might have put together rough sketches and colour ideas to give an idea of what is wanted. In this case the brief will be fairly specific. A company may have seen a range of designs for duvet covers that they like. However, they may be unhappy with the colourways being offered. An in-house designer working for the printer may be asked to re-colour the designs, producing colourways that are considered more suitable.

Perhaps these will have to co-ordinate with a range of lampshades being developed in a specific range of plain shades. A knitwear manufacturer might find themselves approached by a company which has a sweater that they are manufacturing overseas and are currently importing. The company is, however, experiencing real problems with quality and late deliveries. Essentially the company wants the knitwear manufacturer to re-create their sample sweater to a specific price point.

The designer's job in this case would be to source suitable yarn and develop specifications for the production of a garment that is essentially the same as the original sweater in terms of quality and appearance but which also meets the pricing requirements.

A company might have a fabric that is currently selling very well — their 'bestseller'. The marketing director and his team require a 'new'version of this fabric, perhaps more up-to-date and with a co-ordinate. They are happy with the existing colourways but are open to suggestions regarding colour and scale.

As their market area is becoming increasingly competitive they would also like to see the new version having elements of 'added value' if at all possible. This might mean repositioning the product in the marketplace or developing a new product that acts as a co-ordinate with the current best seller. In this case the yarn designer's job is to develop a fabric using the existing colourway that fulfils the company requirements.

THE YARN DESIGN FUNCTION

The Activities of a Yarn Designer

The overall purpose of a yarn designer is to design and produce to an agreed timetable, an agreed number of commercially viable yarn designs. Research into what designers do has identified several different activities. The activities listed below relate to an in-house fabric designer engaged on branded, contract and general work.

- Deciding what to design.
- Producing original design ideas.
- Developing design ideas through to a form suitable for initial sampling.
- Supervising the production of original fabric samples and keeping suitable records.
- Submitting samples to customers, to company or to company selection systems.
- Adapting and modifying designs from sketches or fabrics submitted by customers (or obtained from within the firm or group) to meet a price or other restriction.
- Obtaining the acceptance of a certain proportion of the original sample fabrics by the customer.
- Controlling the production of sample ranges in consultation with the customer's technicians, technologists and buyers.
- Establishing production specifications.
- Ensuring the production of sample ranges by certain agreed dates.
- Reporting to the company on contacts with customers, competitors, exhibitions, developments by fibre and yarn producers, etc., in order to augment market intelligence.
- Controlling the systems and procedures for the storage and retrieval of designs and samples within the company.
- Working within an agreed departmental budget.
- Controlling the supply of materials and equipment for design and sampling.

All yarn designers will certainly be involved in some of these and should have an understanding of all of them.

How Design Work is Done

Design ideas are developed through in different ways, depending upon how an individual designer likes to work, upon the product being designed and upon the manufacturing process/es. From initial paperwork, designers of constructed yarn may take their design ideas and develop them through on the loom or knitting machine. Designers of printed yarn will usually develop their ideas right through on paper. 'Much design work is carried out in a very direct and informal way. The degree of formality becomes a function of scale and the number of interests represented.'

- Design is an investigative process; it involves research. The first stage in any design exercise is normally an enquiry into what the client (or potential client) requires; their needs and expectations.
- Design is a creative process; it involves art and aesthetics. Designs can be copied or invented. Adesign problem is solved with the help

of know-how, ingenuity, pattern recognition abilities, lateral thinking, brainstorming, etc.

- Design is a rational process; it involves logical reasoning in the checking and testing of proposed solutions, information analysis, experimentation, field trials, etc.
- Design is a decision-making process; it involves making value judgements.

The selection of particular combinations and configurations, layouts or shapes involves considerable uncertainties. These are resolved by estimating the values that are likely to be placed on the various major alternatives. In this respect the task of the designer is closer to that of a manager rather than a researcher, scientist or artist. Estimating the social consequences of various design alternatives also falls into this category. Solving design problems involves a mixture of the intuitive (having ideas) and the systematic (rational scientific appraisal). There is no single solution to a design problem, rather there are many solutions although some may be better than others.

The Design Process

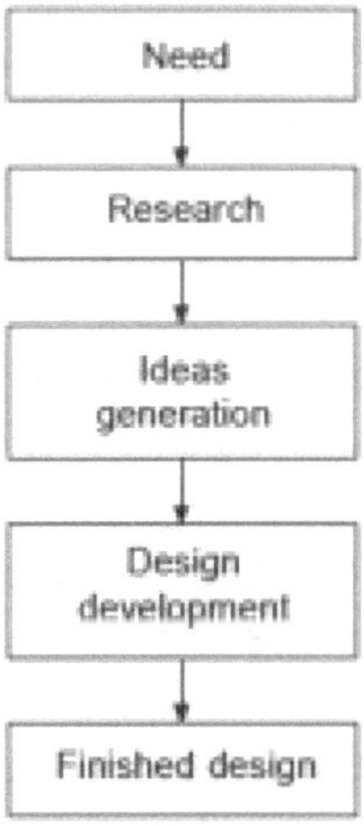

Fig. Need Research Ideas generation Design development Finished design. The design process.

The design process usually starts with a requirement or desire for a new item or product. Research will usually be carried out then to find out as much as possible about this need and about the role or function the new item or product is to have. Ideas generation is the next stage when various alternative initial ideas are conceived. These initial ideas are then usually developed through until the designer is happy to offer them as proposals to meet the initial need. In the early stages, alternative ideas will often also be presented. These proposals will be considered and perhaps modified. A decision is then taken as to the best solution to the design problem and the necessary specifications and instructions will then be given.

Planning Design Work

Any design project will usually need to be completed by a certain date, and more often than not designers will find themselves working on more than one project at any one time. Efficient project management and efficient time management are obviously both desirable. A project entails one or a number of individuals working together over a period of time to achieve an agreed goal or outcome. Any project needs to be completed on time, to a standard and within a budget. A design project, as any other project, needs to be planned and managed efficiently.

The Planning Process

Planning can be described as managing and controlling events to achieve a goal and making the best use of resources. Planning is achieved by making decisions about what is to be achieved (aims) and how best to get there (who does what, when and how).

While planning takes a little time, lack of planning can result in the waste of a lot of time, the expenditure of much money and the generation of considerable stress and tension. A plan gives what could be described as a set of route markers that allow checks to be made as work is undertaken to achieve the goals.

Regular checking of current position against plan identifies any deviations from course which can then be remedied. The more detailed the plan, the more route markers there will be. A plan helps in the controlling of events rather than events determining direction. Very few jobs go exactly according to plan, but if there is no plan there is no way to keep control and track progress.Objectives

An objective describes something to be achieved in the future. The essence of an objective is that it defines what is wanted without describing how this is going to be achieved. Without objectives people do not know where to invest their time or resources, and they can drift in an unco-ordinated way. When objectives are unclear, it is impossible to make effective plans and people will not be committed to action. Despite the obvious importance of objectives many people find it difficult to set them.

Identifying the Aims and Objectives of a Design Project

The aims and objectives of any design project are usually identified at a briefing meeting. This is when the designer finds out what is required of them.

Many clients often have little idea of what they actually do want so it is up to the designer to get as much information as possible; this is usually done by asking questions.

Checklists

Using a checklist can be very helpful at many stages in the design process. For example, a checklist for a briefing meeting would simply be a list of questions, prepared beforehand, that the designer will use as a prompt. Enough knowledge of a situation is required to enable a checklist to be developed that covers the right questions. Use of a properly prepared checklist should ensure that the designer does not forget to ask for any necessary information and this should help avoid misunderstandings that might create problems later in the design process.

Project Planning Methods

Different methods can be used to help plan design projects. The simplest is to arrange elements of the plan in a logical sequence. Working from the brief, designers determine what is required to fulfil this and plan a logical sequence of events to take them through the design process to a design solution that satisfies the brief.

Backwards Planning

Often, the best way to ensure that design work achieves the requirements of the brief is to start by examining the expected final products or outcomes. These products or outcomes need to be listed and agreed. Having considered what it is that is to be achieved, the processes or work to achieve the outcomes can be determined. Finally, it must be decided what inputs are necessary to enable the work to be carried out successfully.

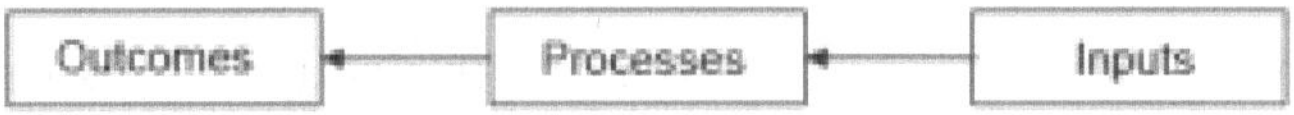

Fig. Backwards planning.

- *Outcomes*: The planned result of the project, equal to the aims and objectives of the project, to achieve the quality or standard required.
- *Processes*: The work needed to be done to achieve the output deliverables or outcomes.
- *Inputs*: The resources in terms of individuals and their skills, time, materials, equipment, techniques, etc., necessary to achieve the outcomes.

For example, a designer is asked to design a range of printed curtaining fabrics. The *outcomes* would be a specified number of fabric designs, with colourways, for a specific time, to sell at a specific price point. The *processes* would be initial research of the market (to see what is currently selling and what the competition are doing), the formulation of initial design ideas, the development of these ideas, the selection and production of the specified number of fabric designs (with recommendations for print base fabrics) to be included in the range, colourway development and selection, and the specification of final fabrics for production. The *inputs* would be the designer with the required skills and knowhow, the time they would require and the materials and equipment necessary.

Gantt Charts

A slightly more sophisticated way of planning is to use a Gantt chart. During a design project, several stages can be under way at the same time. A Gantt chart is a simple horizontal bar chart that graphically displays the time relationships of the stages in a project. Each step is represented by a line or block placed on the chart in the time period in which it is to be undertaken. When completed, a Gantt chart shows the flow of activities in a sequence, as well as those that can be under way at the same time. Gantt charts can also be used to chart actual progress, by drawing lines in different colours to show the start and end dates of each step. This allows easy assessment of whether or not a project is on schedule.

Network Analysis

This is a generic term used for several project planning methods, of which the best known are PERT (Programme Evaluation and Review Technique) and CPA (Critical Path Analysis). These are more sophisticated forms of planning than Gantt charts and are appropriate for projects with many interactive steps.

Project Management

Managing a project involves co-ordinating activities so that they run according to plan. The progress of a project should be monitored and measured against the plan. When deviations occur, corrective action should be taken.

Time management

All projects are time-bounded and for any practising designer it is desirable (if not essential!) to make the best use of the time available. Time management is simply making the best use of time to achieve what is necessary. To effectively manage time, goals and time limits need to be set. What is required? What has to be achieved and when? How efficiently do these goals have to be met? It is only against set targets that success can be measured. The way any individual uses time is unique to that individual. Some people use time as chunks into which they can fit certain activities, all neatly stacked. Others have no clear view of time, selecting activities at random or changing priorities to suit the current crisis.

There are strengths and weaknesses in both ways and it is of value to consider both because many people will alternate between the two, depending on the jobs in hand. The way designers use their time will be different, since everyone has their own pace of work and their own rhythms, with different peaks for different activities. Where possible, work should be done at times to suit an individual's own speed and their own way of organising and completing activities. Managing time costs time. To sit down and plan the best use of time is an investment.

It takes time to learn to use software packages for word processing and database management but their use is a huge investment for future time management. The time spent on any activities should be considered afterwards and evaluated. Can any lessons be learnt to plan more effective use of time in the future? *The Pareto principle: 20 per cent of what you do yields 80 per cent of the results.* Targets that will result in a high pay-off should be identified. Constructive avoidance is when time is spent on work that is neither important nor urgent in preference to urgent and important work. Time management involves:

- *Planning tasks* – getting information, assembling relevant facts, skills, experience, resources, establishing what is known or what needs to be known, processing the information, considering options and identifying the risks involved,
- *Stating what has to be done* – scheduling what will be done, how, where, when and by whom.
- *Getting things done* – doing things and monitoring progress against checkpoints/standards, reviewing outcomes by assessing the results achieved in relation to the aims, determining if more needs to be done, analysing successes and difficulties so as to plan for improvement.

Range Planning

A range is a group of products offered for sale. These products will be the designer's answer to a brief. Having the right products in a range is very important in terms of how well that range will sell.

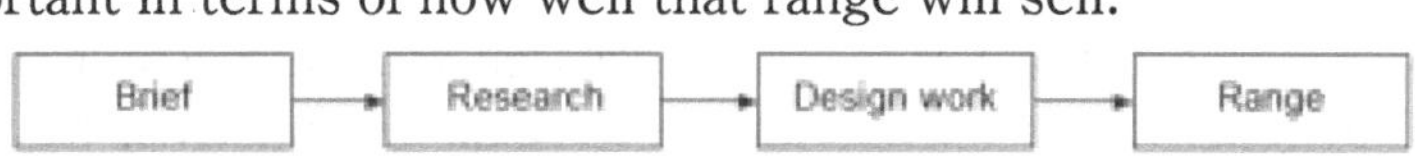

Fig. Brief Research Design work Range. Range development

Good design adds value to a product. With products of equal quality and price, the design will be what differentiates. What does the designer need to know or make decisions about when planning a range? Or, put at its most simplistic, how does a designer decide what to design? All designers are designing for the future and are influenced by trends in their product area. The whole area of forecasting is one where much money is invested to 'get it right' and much money can be lost if a designer gets it wrong. As well as needing to understand forecasting, a designer needs to know other information when planning a range. This includes:

- Type of products,
- Customer/market areas,
- Price points,
- Manufacturing capabilities,
- Quality required,
- Number of designs,
- Number of colourways.

It is usually the job of the design manager (or stylist) in a studio to plan the ranges. There will have been some discussion with sales, marketing and production about all or some of the factors listed above—essentially to build up a full picture of what criteria the range must answer.

The type of designs depends on the products being designed for, the customer profile, price points for sales, manufacturing plant availability and quality.

The number of designs and colourways in a range is controlled by several commercial factors including the amount of capital the company wishes to tie up in the range, the raw materials that are needed, the amount of stock that can be carried, and the production capacity, *i.e.* the number of machines available for printing/knitting/weaving the range (some machines may be tied up with specific customer orders rather than general range production).

Research

After the briefing meeting, the next stage in the design process is usually research: finding information about products and markets, production processes and techniques, etc. Information can be collected in different ways:

- By waiting to see if information presents itself (gathered by chance from contacts and perhaps by reading). This is unsatisfactory if there is any form of time limit on when the answer is required.
- By asking questions. This is appropriate if it is known who to ask and what to ask. Known as primary research or field research, this is where designers find information directly for themselves; by going out and looking at the shops and visiting trade fairs and exhibitions, by asking relevant individuals and groups, by using mailed questionnaires, by setting up focus groups and by interviewing people.

- By looking up what is required. This is appropriate if it is known where to look, how to look it up and what exactly is being looked for. Known as secondary or desk research, this is where information is gleaned from work already carried out by other individuals or groups. Such research might include looking at market research information obtained through government censuses and large market research programmes.

Information Search

To ensure that the required information is found and time is not wasted, the collection of information should be systematic. The aim should be to find key information. Identify what is not known.

As far as possible, vague requests for 'information on' some problem area should be broken down into questions defining what it is that is not known and what needs to be found out. The questions that need to be answered should be identified and listed. The level of answer required should be established (there is a balance between the importance of what is being looked for and the time and effort spent looking) and a checklist of likely sources of reliable answers to questions can be prepared.

It is sensible to keep control over any search for information by setting limits and deadlines and by using expert opinions and expert literature searchers to identify the most promising sources. It is important to stop as soon as there is enough information to move forward and it is essential to keep accurate references.

Getting Information From Other Organizations

There are many occasions when designers need to find out information from other organisations. Sometimes, the information sought may have to be bought, sometimes it may be free—part of a marketing strategy on behalf of the supplier.

However, whether visiting a new supplier to source yarn or buttons, or telephoning to check prices of fabrics, it is important that designers know exactly what they want so as not to waste time. Before any visit, the information required should be clearly identified.

It is essential for any designer to be prepared, and establishing a checklist of questions that need to be answered is a good way of making sure that all relevant information is obtained.

When seeking information about market areas and products, it would be usual to employ professional market researchers. However, sometimes designers may prefer to carry out some market research themselves. In either case, well-constructed questionnaires can be very useful, and when market research is being carried out by professionals the designer's input should not be overlooked.

Questionnaires

Questionnaires should be designed in such a way that they take a minimum amount of time to fill in. A busy company director or buyer is much more likely to complete and return a questionnaire if it is clear to read, to the point and does not take up too much of their valuable time. Formulating questions in such a way that they can be answered by ticking boxes is much more likely to elicit a positive response than having questions that require written replies.

Ideas generation

With some initial research undertaken, the designers need to be able to generate suitable ideas to help them answer the brief. Design ideas can come from almost any source, but sometimes designers may find difficulty in thinking of suitable sources and themes for work.

Brainstorming

Brainstorming is a very useful method of generating ideas and can be used at many stages in the design process. In its most formal sense, it is a group participation technique for generating a wide range of ideas in order to tackle a stated problem. In a less formalised way, an individual or pair can use a brainstorming session to generate ideas. What follows is a description of brainstorming at its most formalised; such a rigid approach, however, may well inhibit ideas.

- A problem statement is formulated. Too vague or too restrictive a statement should be avoided.
- The group of people to participate in the session is selected. The group should ideally include some people familiar with the problem area. Small groups of about 4–8 people are best.
- Five or ten minutes are allowed for group members to write down their first ideas in reply to the problem statement.
- Group members are encouraged to continue writing down new ideas whilst each person in turn reads out one idea from their set. The session rules are:
  - — No criticism is allowed of any idea.
  - — Crazy ideas are welcome.
  - — The more ideas the better.
  - — Ideas should be combined and built upon.
- After the session, the ideas are evaluated.

Range development

So how does the design manager go about putting a new range together? Once a basic colour palette has been decided, this will help the ideas being developed to work together as a total package. Colour co-ordination does not mean that every design has to be in exactly the same colourway, but colours should be related.

New fashion colours should be looked at alongside any existing palette to make sure that these work as part of the range. Decisions have to be made as to whether best- selling designs from a previous range are going to be carried on. If a design was well received and is still selling well, it makes commercial sense to carry it on. It is hard to argue with sales figures. When continuing a design through to the next season, it must be decided whether to change the colourways to bring it in line with the new season's colours or whether to continue the design in the existing colourways and add some new colourways.

Re-colouring can add to the sales of a proven design. Areas and themes will have been identified according to the customer requirements, as will the appropriate number of products to fall within them. Having established the number of products and colourways for each, initial ideas for the range can be put together. At this development stage it is very necessary to maintain an overview of the whole range.

If it is a range of apparel fabrics that is being developed, then there might be part of the range aimed at menswear and part at womenswear. An important existing customer may be a pyjama manufacturer and so designs suitable for that customer must be included. Designs may be developed in-house or bought in, or a mixture of both.

Designs may come from archives of designs that have already been developed and sampled. A few designs can be assembled around each theme or group and any areas where it is felt designs are missing will prompt more sampling.

Production capabilities and machine loading must be considered. For example, there is no point when designing a range of knitted fabrics to produce all the fabrics for this on one specific machine type if the manufacturing company owns several different types of machines, producing quite different fabrics. It is only when all such considerations have been taken into account that the final selection of the designs to be included in the range will be made. Once fabric designs are chosen, suitable colourways will be selected, and if necessary, sampled. Colourways may be balanced or not. Balanced colourways are when the colours change but the relationships of the colours within the design stay the same.

Unbalanced colourways are when there are no similar colour relationships between colourways. (For example, a three-colour print design has a dark blue ground, large flowers in midblue and small flowers in light blue. A balanced colourway of this would be one that had a dark green ground, large flowers in mid-green and small flowers in light green. An unbalanced colourway would be one that had a bright yellow ground, pink large flowers and dark blue small flowers.) For a range to work as a total package, themes and concepts can also be used as well as colour and colour co-ordination.

For example, Dorma, a UK bedding manufacturer, produced a very successful range of bedlinen based on the illustrations in the book *The Country Diary of an Edwardian Lady*. This range was so successful for Dorma that several seasons later a further bedding range was developed from the same book, but this time using a different colour story. Themes can be used as a base for designs and colour co-ordination. Themes and concepts can also help with publicity, being taken through to brochures and in-store presentations.

Many fabric ranges will include plain colours. In most product ranges there are basics that have to continue, and decisions have to be made as to which colours will be carried on and which will be dropped. Often, these plain colours are related to other products in the same area; for example, towels relate to bathroom fittings, telephones to paints and wall coverings. There should be some co-ordination between plain and patterned ranges. This allows customers to buy from both and thereby increases sales. Range planning and development is very important. The brand range is what a company sells itself on. The image created with the range will be the image of the company for the next twelve months or so until the next range.

Design Development

The way design ideas are developed will depend on two main factors; how an individual designer prefers to work and what is being designed for. Designers will usually try out different ideas, taking the best of these through into fabric or, in the case of print design, through to designs painted-out to size and in repeat. Computer-aided design (CAD) systems can be very helpful at the sampling stage, allowing different ideas to be tried and fabrics to be simulated without the expense of fabric sampling.

Sampling

Fabric sampling is an expensive process. Very often, machinery has to be taken out of production for sampling, and time that is not being used to make saleable products on expensive machinery is costly. Money tied up in sample stocks is also money that cannot be recouped.

Range Presentation

The introduction of new products to a company's product range used to be done in a very arbitrary way, with little planning or co-ordination. In the early 1970s however, things began to change. Many companies closed due to world-wide recession. Those that survived had to work harder to maintain a place in the market. Ranges had to be well presented, with marketing becoming increasingly important.

For example, for yarn manufacturers and spinners it was no longer sufficient to show new yarn just on cones. Ranges had to be carefully planned, the presentation of the range to the customer became very important and the presentations began to include how a range might be used. Spinners started to employ designers to put together trend ideas. Different yarn qualities within a range would often be colour co-ordinated to help the customer maximise use, and fabrics that illustrated the current trends would be made up from the yarn in the range.

Presentations of a range are often both written and oral. There will normally also be drawings and photographs illustrating how fabrics from the range work together, how fabrics work when made up into or incorporated into products and how these products work with other, related products. Presentations often reflect the theme used for the design work. For a presentation to be successful, the designer must have answered the brief. If the brief is not clear, it is difficult to solve the design problem. At a briefing meeting, the designer is trying to get as much information as possible. Everyone who briefs a designer will have a preconceived idea of how a job will turn out and the designer's job is to find out what this is. It is therefore important that presentations should be made to the person who gave the briefing. Presentation of the final range to the person who asked for the work to be undertaken is, however, not the only type of presentation a designer will be called upon to contribute to or to make.

Once the range has the approval of the relevant directors or company board members, the designer may well be asked to make presentations to sales teams, agents and customers. Designers are called upon all the time to give presentations of their work and so an ability to present themselves and their work well is very important.

Presentation of Initial Design Ideas

At various stages of the design process, designers are often called upon to present their ideas to their colleagues, managers and directors. Designers have to look the part, they must present a 'designerly' air and should show themselves as being organised and capable. The best presentations are simple and to the point; this is as true of visual presentations as of written and oral presentations.

Presentation of Design and Artwork

Designers are visual people. Their work will often need little verbal explanation, clearly communicating the intention by virtue of visual impact. The way such work is presented is very important because good presentation can show a design solution to advantage while poor presentation can hide and distract from good design. Presentation is itself an exercise in design. Good presentation should be suitable and professional. It should be suitable in that the presentation style and techniques should show whatever is being shown (be it fabrics, yarn, artwork or drawings) to its best advantage; it should look as good as it possibly can. All presentations should be professional in that they should look neat and tidy, deliberate, considered, planned, well designed, appropriate, consistent, related and reflect the work presented. Poor presentation is chaotic, messy, dirty, disorganised and inappropriate.

Visual presentations made by yarn designers

The types of visual presentation of their work that yarn designers find themselves making are for a variety of purposes:

Mood/theme Boards

These boards are collections of ideas around a mood or theme. The ideas may come from magazines, postcards, colour swatches, yarn or fabric pieces. Presentation of mood boards should be simple and suitable; the style of presentation should reflect the mood. A sporty mood board will be presented differently from a mood board reflecting luxury and wealth.

Initial design ideas

It is often appropriate that designers themselves present their initial design ideas to a customer or client. This will allow the client to feel that they have some input and also should prevent unsuitable work being done. The client may be in-house or external. An external presentation may be more formal than an in-house presentation, but the presentation should in both cases be appropriate and show the ideas to their best advantage.

Finished design work as the solution to a design problem

Finished design work is presented to show the design solution to the brief. It will often show the development of ideas through to the final design work.

Work to show capabilities and potential

Designers will build up a portfolio of art/design work to show what they are capable of. Art/design work should show a variety, a selection of the best work showing how ideas are worked through. It should be neat and tidy, and presented in a coherent manner. It should be easy to look at. When making decisions about what should be included in a portfolio, it is necessary to consider who will be looking at the portfolio—how much time will they have? What will they expect to see? etc.

Publicity/promotion/packaging

If a range is to sell well, it must be presented in a simple and understandable way to the non-designer. Publicity can contain photographs of furnishing products in room settings, emphasising the importance of colour co-ordination and pattern co-ordination. It can be seen how the range is related to other products in the same room area. Publicity for fashion garments will show how the garments in the range can be used with other garments. If the range has a name or theme this can be emphasised, for example through publicity with a distinctive logo.

A logo can be used on packaging and on any labels. If the product is not fully visible when packaged, then a photograph will be needed. This technique is often used for duvet packaging; the folded duvet will be packaged with a photograph of the duvet *in situ*, allowing an appreciation of the full design. Such photographs may well also be used for in-store displays. If the product is targeted as a gift then the packaging should be especially attractive. Soap may be sold with a towel. Products sold in a basket mean that the packaging becomes a product in its own right.

Store display

The days have gone when, in a department store, the bathroom fittings were in the basement and towels on the fifth floor, with no hint of a relationship. Concept selling offers many related products in an area, and bedding, curtains, lampshades and other items for the bedroom may be displayed together, co-ordinated by colour and a common or linked theme. Such selling techniques make it easier for the designer to get across the message of co-ordination and for the customers to see how the product will work in their home. However, presenting merchandise in this way is not always easy for department stores, where buying systems have been traditionally broken down into product areas.

Presentation to customers

This is when the designer shows the finished range of products. The presentation may be to selectors from a major high street store group, or it may be to customers from within the designer's own company or group of companies. The designs may be either the work of the designer doing the presentation or the result of several designers and others, such as technicians and sales personnel, who have all had an input. Design is very much a team activity with the designer co-ordinating the design development, whether this is for one fabric or a range of fabrics. Designs have to fit into a customer's market area so designers need to be familiar with their customer's current ranges.

They have to try to identify gaps in product and colour ranges and their ultimate design work should be to fill these gaps. Designers need to try to identify new directions for their customers. It is important that new ranges lead the way rather than follow so that customers come back for new ideas. Storyboards are frequently used to illustrate points in such presentations. Feedback from the customer on designs and directions should be encouraged to ensure that the customer is involved and has a degree of ownership, and that the brief is answered. This also helps establish good working relationships.

THE PRINCIPLES AND ELEMENTS OF YARN DESIGN

Yarn are frequently made to be decorative and are used to embellish and decorate both people and objects. The designers responsible for such yarn have to balance many factors when answering a design brief. What is the fabric for? How must it perform? Who is the customer? What are the economic restraints? How is the fabric to be produced?, etc. Research into how people select products shows that colour and appearance are two of the most significant factors, with handle, performance and price coming lower down in terms of importance. Yarn designers therefore need to have a good understanding and sensitivity to colour and aesthetics. The same design elements and principles apply to design as to fine art, and an understanding of and an ability to use these are as vital to the designer as to the fine artist.

'Design is about relating elements, whether they are similar or contrasting and visually arranging an interesting unity with them. Shapes, forms, colours and texture all combine to become a unified whole which is commonly called a design.' Art and the act of creating are highly personal activities. In the first instance it may seem impossible to objectively analyse works of art. In different cultures and at different times, styles of art and even the media used to express art have varied enormously. There are, however, certain basic design features that appear to be natural and common to all art.

Design Principles and Elements

Design features, the elements and principles of design, can be called the language of art and design. Pick up one of the many books on design and these elements and principles will be there, although they may be detailed slightly differently in each. This is, in part, due to the way that in any one work of art no one element or principle can exist in isolation. Designing can be defined as relating and visually arranging components or elements to create effects. Space, line, shape, form, colour, value and texture are the design elements with which artists and designers work to create a design. The design principles of balance, movement, repetition, emphasis, contrast and unity are what artists and designers do with the design elements to make the art form or design.

Design Elements

Line

A line can be a mark made by a pen or drawing instrument or it can be any continuous mark that causes the eye to follow along its path. The viewer's eye travels along the path of a line because a line is longer than it is wide. A line moves, and as it does so it indicates direction. A straight line leads the eye swiftly across the picture plane but the eye travels more slowly when following the path of a jagged line. Lines appear in different ways. There are curved lines and straight lines.

These can be long or short, thick or thin, ragged, sharp, light, dark, simple or complex. Lines can be broken and yet have direction. Lines can be textured and can be coloured. Lines can be made from repeating similar elements, as diverse as dots or people, in a lengthways direction. Lines can be used to create form, to give depth.

Lines can be carefully controlled to create optical sensations and can be used to project feelings of sensitivity and strength. When someone looks at lines they try to fit them into something related to their previous experiences; for example, scribbled spirals may be interpreted as seashells, and a few lines can easily suggest an apple to someone who is familiar with this fruit. In some ways this can restrict the artist's or designer's freedom to make viewers see what is intended, although it does also open up new possibilities. Skilful artists and designers play with this tendency to see familiar forms in everything by using only a minimum of visual clues to evoke a far more elaborate response. Care must be taken, however, that abstract designs and patterns do not unintentionally suggest unpleasant or ugly forms to the viewer.

Shape

When a line turns and meets up with its start point, a shape is created. A shape or figure is a positive thing and occupies positive space. The area surrounding a shape is called the background or ground. It is a negative thing and occupies negative space. Shapes can be clearly defined with hard edges but often they are not clearly defined, which means that it is more difficult to see where shape ends and background begins. There are a great variety of shapes to be found in nature. Many artists have drawn creative inspiration from natural shapes. Through their imagination, artists have invented new ways to use shape and communicate ideas. Shapes can be solid or opaque, linear, textured, coloured and outlined. Shapes can be transparent, revealing other shapes behind them. Similar shapes need not be identical, yet they can have a common relationship, which visually ties them together. Some shapes will command more attention than others, depending on their size, colour, value, texture, detail or their location in relation to other shapes. Tall shapes are elevating, long flat shapes express calmness, downwardpointing shapes activate the sense of falling.

Form

Form is structure. In art and design, form is the illusion of three-dimensional volume or mass seen in two dimensions. Careful observation of the forms around reveals that, in nature and in man-made objects, many forms can be described as combinations of the basic geometric structures; spheres, cylinders, cones, cubes and pyramids.

Paul Cézanne, the impressionist artist, in his landscapes and still lifes, explored how these basic geometric shapes combined to make more complex shapes. There are several methods that can be used to give an illusion of three-dimensional form and some of these are detailed below. Line, shape, colour, value and texture can all be used to suggest form by the artist on a picture plane, and by the yarn designer within fabric.

Space

Space flows in, around and between forms or shapes. A flat surface has only two-dimensional space. This means that it has length and width but no depth. It is impossible to create actual depth or space on a flat surface but an illusion of space, distance or depth is possible. There are many methods used by artists and designers to create this sense of space, and to convince the observer that there is space and depth when, in fact, they are victims of a type of visual deception. Objects placed higher up can create the feeling of depth or distance.

Overlapping shapes can also create the feeling of depth, as can converging lines. If, in two-dimensional art and design, space exists purely as an idea or concept, it leaves the artist or designer free to compress or stretch it as they wish, to portray the particular feeling that is desired. Colours have an effect on space. Colours that are warm and bright appear closer, while dull or cool colours recede into the distance. A flat surface that is covered only with pattern can eliminate any feeling of space.

Colour

Under normal light conditions, well over 10 million different colours can be seen. Throughout history, the study of colour and the development of colour theories were frequently undertaken by artists and designers, and many artists have spent their lives attempting to understand colour. Seurat wanted to apply a scientific system to the methods of the Impressionists, and his Pointillist theories, where dots of colour from a restricted colour palette were used to create the impression of a wide range of colour, clearly show his fascination with analysing colours.

Colours communicate; some colours are associated with cold (blues and greys) and some with warmth (reds and oranges). Colour can convey the time of day, weather conditions and temperature, and even the time of year. Colours can be designed to blend in with the environment or to stand out. Frank Lloyd Wright saw architecture as growing directly from the earth on which it stood and therefore used the colours of the surrounding areas in his buildings while Richard Rogers and Renzo Piano, the designers of Paris's Pompidou Centre, used colour, not to blend in, but to code the different pipe systems for heating and other services. In art and design work, the colour can be used in a natural or abstract way.

Maps make use of both abstract and natural colour; natural colours are used when illustrating mountains, deserts and seas while maps showing political boundaries will use abstract colour. Colour is a crucial part of all branches of design and design-based industries. Advertising agencies know that graphic information in colour will have a more profound effect than that in black and white. Colour grabs the attention span because memory recall from colour is quite pronounced. Colour is used to promote corporate identity. Blue is used by many banks to give the suggestion of reliability, while the appetite colours of red and yellow are used by fast food chains. Green is frequently used to denote environmental friendliness while more subtle greens convey a feeling of upmarket status.

The *American Express* card was originally launched in yellow but was a flop; however, when its colour was changed to that of the American dollar bill it became a success. The green livery of *Harrods*, the world-famous upmarket store in London, epitomises good taste and sophistication. Forecasting colour trends is itself an industry, and fashion colours change with the seasons. Underlying these cyclical trends, however, are some basic colour preferences. In colour popularity tests, blue is frequently placed first, with red second. Blue ties are most popular, with red ties in second place, and blue and red cars regularly occupy top positions, only being outdone by silver, a colour symbolising luxury and wealth.

The colour of a product influences the perception of it, and this is used extensively in marketing. Many people believe a red car will drive faster than a white one. Tests with coffee showed that coffee served in a red mug was preferred to the same coffee served in a yellow mug (considered too weak) and in a brown mug (too strong). Foodstuffs claiming to be pure and unadulterated often use blue and white packaging to communicate purity. However, in this age of environmental awareness it is predicted that more earthy colours will symbolise a natural product while white will be associated with chlorine and all that is environmentally unfriendly.

Products communicating strength adopt vibrant and contrasting colours; the greater the contrast, the stronger the associated power. Colour can be used to represent a product, with some colours idealising a product, and some using biological signals to communicate function. Yellow and black are used in nature as a warning symbol for reptiles and insects that have poisonous bites or stings and this colour combination is therefore often used to represent danger. Yellow and black is used on signs when the desire is to signal caution.

Colour mixing

There are two ways to mix colour—'additive' and 'subtractive'. Additive colour is the mixture of coloured lights. The three primary colours of red, green and blue when mixed together in equal colours produce white light. Mixing the three additive primaries in differing amounts of coloured light can create any colour in the rainbow. Colour televisions use the principle of additive colour mixing. In subtractive colour mixing the principle is exactly the opposite. The subtractive primary colours are cyan, yellow and magenta. When mixed together they subtract from the light producing black. When different pairs of the subtractive primaries are mixed, the colours red, green and blue are produced. These principles are used when we paint and in photography and colour printing.

Colour systems [ssssh]

As the human eye can distinguish between 10 million colours, it is clear that to describe colour experiences by name is imprecise. While everyone knows what is meant by tomato red, different people have different ideas as to exactly what colours are meant by beige and sand; mauve and lilac. In order to accurately describe or pinpoint colours, a reference or colour system is required. All the attempts to notate colour can be traced back to the work of Sir Isaac Newton. In 1660 he re-created a spectrum by directing a narrow beam of white light through a prism; he went on to develop a colour wheel by taking the two ends and bending the spectrum into a circle. This colour wheel evolved and changed over the centuries and in 1810 Otto Runge created a spherical model, with white at the North Pole and black at the South Pole and with Newton's colour circle forming its equator.

In 1915 Williem Ostwald devised his double-cone colour solid; also in 1915, Albert Munsell developed another system of colour notation that added steps to the constituent hues. Munsell allowed his three-dimensional colour solid to respond in shape to the different potential strengths between hues, creating an asymmetrical colour solid. His colour model can be explained in three dimensions, in terms of hue, value and chroma.

Hue is the 'colour' of a colour, *i.e.* its redness, greenness or yellowness, *value* refers to the amount of lightness or darkness, while *chroma* refers to its saturation or its colour strength. Anumber of commercial colour specification ranges have been developed from Munsell's work including the BS4800 and the Pantone and Colour Dimension systems. In yarn and clothing, the *Pantone* system is probably the most commonly used, and a whole range of products is available in its defined colours. These include ranges of marker pens, paper and even fabric samples.

Colour Psychology [ssssh]

There is a commonly held belief that red is exciting and green calming. But which red and which green? Research has shown that strong greens, yellows and blues are all seen to be as exciting as strong red. The same research also showed that pale and dark versions of the same hues have quite the reverse effect. Pink has a calming effect on violent people.

A common belief is that blue increases the size of a room while red has the opposite effect. Recent research, however, has shown that it is not the colour itself that has this effect but rather its value, the amount of whiteness or blackness. However, with regard to temperature, colour does have a very definite effect. Red and yellow rooms are perceived to be warmer than blue and green rooms.

Value (tone)

Between the whitest white and the blackest black there are countless degrees of light and dark values. Value is important to the designer. Shapes that are close in value or tone appear to merge together. Visually dark values appear to come forward and light values tend to recede, but the reverse can occur. Sharply contrasting values attract attention and the use of light against dark or dark against light can create the illusion of size difference. Value is probably the most elusive of all the design elements. The success or failure of a piece of work may rely on the use of values within it.

Texture

One definition of texture is the surface of a material as perceived by touch. Running a hand over a surface may find it to be smooth, rough, dull, glossy, hairy, sandy or bumpy. Texture is very important in yarn, and how fabric feels is an important consideration in product choice. Texture can be seen as well as touched and artists and designers can use a variety of techniques to convey textures when none is actually there. Repetition of design elements can often create a visual illusion of texture.

Design Principles

In every design or work of art there are some or all of the design elements—varieties of line, positive and negative shape, three-dimensional form, occupied and unoccupied space, colour, value and texture. The manner in which these elements are used and combined determines the quality of a piece of work. Thoughtfully balancing, moving, repeating, emphasising and contrasting the design elements can achieve a unified piece of artwork or design.

Balance

Balance is a sense of stability when applied to opposing visual attractions or forces. There is a natural desire for balance, and in nature balance is ever-present. In formal balance, the design elements are almost equally distributed. A design or composition that is divided in half so that one side is the mirror image of the other is said to have symmetrical balance.

In radial balance, the design elements radiate from a central point as the spokes of a wheel or the natural form of a daisy. Asymmetry uses informal balance; a centre line or point is ignored, with the design elements being balanced visually, rather than in a symmetrical manner. The position of any particular shape in a composition contributes to its strength. A shape in the exact centre of a picture plane is at perfect equilibrium. Moving the same shape off-centre can increase or decrease its importance. In any yarn design, each shape affects everything else. Exciting balance or imbalance is usually only achieved after frequent arranging and rearranging.

The study of composition has been a fascination for many artists and designers. The Pointillist artist Seurat was as equally absorbed in the study of composition as in the study of light.

Movement

Movement and the portrayal of movement have always fascinated artists and designers. By careful arrangement of the design elements, the illusion of movement can be created. In optical art and designs the sensation of movement may deeply affect the viewer's responses. Some paintings can provoke dizziness by making it difficult for the eyes to focus on a central point. Associated with movement is time. Pictures and patterns are capable of holding our attention for varying amounts of time. Some designs may be so subtle that these are barely noticed, if at all, by the viewer, while others can hold attention for much longer periods.

Repetition

Repetition occurs when elements that have something in common are repeated. When a design consists of shapes that are exactly alike, repeated in a uniform and regular manner, then that design tends to seem more formal. By varying the shapes and the spaces between them, a more informal interest is created.

The repetition of some of the elements within a design repeat can hold designs together. Repeated shapes make patterns. Many yarn designs, because of the method of manufacture, will automatically repeat.

Emphasis/contrast

Repetition occurs when elements that have something in common are repeated. When a design consists of shapes that are exactly alike, repeated in a uniform and regular manner, then that design tends to seem more formal. By varying the shapes and the spaces between them, a more informal interest is created.

The repetition of some of the elements within a design repeat can hold designs together. Repeated shapes make patterns. Many yarn designs, because of the method of manufacture, will automatically repeat.

Emphasis/contrast

Unity exists when all the elements in a design work together harmoniously. In a unified design, each element plays an equally important part.

Inspiration for Yarn Designs

Inspiration for yarn designs can come from a variety of sources. It is possible to create yarn design work by going straight to the process being used. For example, a weave designer may go straight to the weaving shed, choose warp yarn in colours that appeal and, with a knowledge of fabric structures, make an attractive fabric.

Experience, however, has shown that such a method of designing often results in fairly mundane design. Innovative, exciting yarn design starts with much more fundamental paperwork; drawings and paintings exploring colours, textures, shapes and patterns.

Drawing from objects in an imaginative and open way will inspire new colour combinations, textural ideas, shapes and arrangements of these. Inspiration for such paperwork, and ultimately for yarn designs, can come from many things.

Both natural and man-made objects can inspire. Whether the source material is natural or man-made, abstract or tangible, how themes and source material are used is a matter for debate. Drawing from man-made items such as architecture can raise the question as to whether or not the resulting design work is truly original. Drawing from architectural patterns can be seen, if the work is very literal, as merely copying someone else's designs — the proportions and shapes have been created by another person. Working from nature, however, is usually seen as producing more original work. The designer may still be copying but is constantly making decisions and judgements about what to include and what to omit.

Being influenced by the work of artists and other designers raises the question— when does a design influenced this way become a copy? Is a yarn design that lifts elements from other designers or fine artists good or bad design? Fine art can inspire yarn; part of a menswear collection by the fashion designer Paul Smith in the late eighties was made in fabrics that were inspired by the paintings of Matisse.

Yarn can be created by fine artists; tapestries and silk scarves have been designed by artists and sculptors such as Picasso and Henry Moore. Yarn can have fine art directly applied; reproductions of the Mona Lisa have been applied to a variety of products, from tea towels to tee shirts. Much of yarn design is derivative. Source books for designers often consist of designs from the past or from other cultures.

William Morris, the famous nineteenthcentury writer and designer, often based his designs on Persian yarn designs and many of the designs credited to him that are still best-sellers today were almost literal copies of the original Persian ones. Does this knowledge change how Morris's work is perceived? Does using archive or historical sources allow copying to become acceptable? Whatever the inspiration or source for a yarn design, the yarn designer needs to have a good understanding of colour and aesthetics. A yarn designer must be able to draw. They must understand and be able to use colour and pattern.

Pattern

Any yarn designer needs to have a good idea of how their design is going to look as a finished fabric. One piece of paperwork could be worked through to become many different finished designs, woven, knitted or printed. All designers need to understand how designs can repeat.

While many designs work as straight repeats, where the design element is repeated again and again, one on top of another, side by side, there are also half-drop repeats and tile repeats. Mirroring, where elements are reflected about either, or both, x and y axes, is another common way of repeating elements.

Fabrics may have sections within one complete repeat that show other repeat patterns, while some fabrics such as border designs will employ repeated designs only within certain sections of the fabric. It may be useful to mask-off certain areas of paperwork (or sample blankets and sample fabrics in the case of woven and knitted fabrics respectively), or use a frame to establish the most suitable sections to develop.

It is important to consider the repeat structure from an early stage in the design development, and various sketch plans of possible repeat structures should be evolved. Very often, trying to interpret paperwork too literally can be a problem, particularly for a print designer. By its very nature, weaving and knitting design requires elements to be extracted from the initial inspirational paperwork. For print designers, this decisionmaking process of what to include and what to discard is equally important. The best designs are so often simple. This advice is as true for printed yarn design as for any other type of design. Everything that does not contribute, everything that is not necessary, should be left out.

9

Yarn Production of Tradition, Modernity, and a New Middle Class

Perceived as both a lure and a threat, modernity was experienced by many I grew to know in India in the early 2000s as a time, indeed, when "all that is solid melts into air." Such perceptions of modernity were intimately tied as well to visions of the past; for as scholars have long pointed out, the project of fashioning modernity is always also one of fashioning a past that is now lost or pushed away.

A s we were winding our way through heavy traffic one warm Kolkata evening, I mentioned to my driver Sharfuddin Ahmad that the next day I was pleased to be visiting the daughter of a married couple living in an old age home whom I had come to know well. I remarked that I was very interested in meeting the daughter, as I had barely ever seen any sign of the children of old-age-home residents. The daughter has a brother, I mentioned, so maybe I'll be able to find out from her why her brother doesn't look after his parents? My driver, a middle-aged man of four children approaching their marriageable years, promptly and pensively replied:

- Oh, madam, these days children have become such a way that they just don't look after their parents. This age is very bad, you know? If old parents have some money, *then* the children will stay with them and look after them, as long as the parents are advantageous to them financially. But if they don't have money, after they retire, the children will just push them aside. After you give a son's marriage, you never know—What will the wife-daughterin- law be like? Will she or will she not want to look after her mother-in-law? Both those who have love marriages, and those whose parents arrange their marriages—in either case, there's no certainty what the wife will be like. And the younger generation these days prefers to live separately. I've seen movies and things about the youth of these days—what has gotten into their heads, I just can't understand! Those who have

> money put their parents into some kind of an old age home, and those who don't just push their parents out somewhere into the street.

Kerala film director Jayaraj's award-winning movie, *Pathos*, portrays an aging Kerala couple, abandoned by their children who have settled in America. The couple eagerly plans for a visit from their sons, cleaning the house, preparing food, and putting up a swing in the garden for the grandchildren. Then the news comes: the children have canceled in favour of a trip to the Niagara Falls. Worse yet: the sons arrange to sell the ancestral property and place their parents in an old age home.

Director Jayaraj comments on his motivations in making the film: "In Europe, it may be normal that children leave home. But in our society, we have roots, and suddenly, all these families have started sending their children abroad; the children lose contact with their past; they forget to come home." Critiquing the contemporary materialism that has supplanted family intimacy, he adds: "These children who have settled in America are not poor. They make a lot of money and they just want to make more money". The director reflects in another interview: "Even in a small province such as the one I live in, the number of old age homes are increasing every day. Glorified mortuaries with dead bodies of parents wait indefinitely for their children, settled abroad.

These real-life situations motivated me to make this film". On the occasion of the International Day of Older Persons on October 1, 2004, Government of India ministers published messages in newspapers around India. "We, the Indians, belonging to a glorious cultural tradition, have the inbuilt quality of respecting all the elders," begins a statement by the Minister of State for Social Justice and Empowerment.

"We are the pioneers in the field of civilization and the rays of civilization have radiated throughout the world from the heart of India," the letter goes on. Yet, the Minister cautions that "recent times" have witnessed an "erosion in the value system" that may "lead to a colossal disaster." The letter closes: "Let us make the light of love ever-shine in the evening of the lives of our elders. Let there not be any sunset in their lives." Prime Minister Manmohan Singh proclaims in his accompanying message: "We must . . . work to harness our traditional values of respect and reverence for the elders with existing policies." A s is clear from these three brief vignettes, many frame their talk of contemporary aging against the backdrop of a contrast with a traditional Indian past.

The ways people conceive of their past, experience the present and imagine the future are intricately wrapped together. I turn, then, to look more closely at my informants' and public media constructions of the past, as an important context for understanding the ways people are working out modes of aging in the present.

FASHIONING THE PAST

As I analyze narratives of aging in a traditional past, my aim is neither to claim nor to disclaim that this past "really was." When people speak about the past, they are always selecting certain elements from their recollections or perceptions, while neglecting to invoke others. What people construct or experience as past "tradition" *now,* only also becomes tradition in contrast to and in the present context of modernity. So, tradition itself, like modernity, is not something that is fixed and outside history and contemporary human agency, but rather something that is being actively constructed, interpreted and used in the present.

Aging in the Joint Family: Intergenerational Reciprocity and Seva

In contemporary narratives, it is the multigenerational joint family more than anything else that represents tradition in contrast to an emerging modernity. Although a "joint family" technically refers to a family consisting of two or more brothers living together along with their parents, spouses and children, the phrase is more loosely and popularly used in India to refer to any multigenerational household including at least one senior parent and one married adult child with spouse.

In a joint family system, old age itself is essentially a family matter, and adult children, in particular sons and daughters-in-law, live with and care for their aging parents—out of love, a deep respect for elders, and a profound sense of moral, even spiritual, duty to attempt to repay the inerasable debts they owe their parents for all the effort, expense and affection the parents expended to produce and raise them. Such joint family living pertains in contemporary narratives not only to care in old age, but also to a much wider host of social values and meanings, including supportive interdependence, fellow-feeling, a time when kinship was more important than material success, warm rambling households, connection to village lands, moral-spiritual order, patriarchy, tradition, and "Indianness." A t the heart of the joint family is a system of intergenerational reciprocity. Significantly, people indicate that it is precisely what parents once gave their children—such as a body in birth, food, material goods, money, a home, forms of love, the cleaning of urine and bowel movements—that children are expected to provide in return for their parents, years later in old age and by reconstructing and venerating them as ancestors after death.

Although the common discourse is that *children* provide for and co-reside with their senior parents in these ways, in practice it is *sons* and daughters-in-law who are expected to fill this role. Upon marriage, daughters formally relinquish obligations to their own parents, taking on responsibility for their in-laws.

The providing of care for seniors in the family is often termed *seva,* service to and respect for the aged. *Seva* can be offered to deities as well as elders. When provided to elders, *seva* entails acts such as serving food and tea, massaging tired limbs, combing hair, bringing warm bath water, and offering loving respect—in short, striving to fulfill all of the elder's bodily and emotional needs. Related to *seva* is the familiar gesture of *pranam* practiced by many Indians, when a junior bows down to touch an elder's feet in a sign of respect, and in turn the elder places his or her hands affectionately on the junior's head and offers blessings, such as "May you live well," "May your children be well." Juniors perform *seva* not simply as a gift in the present, but in exchange for the elders' earlier tremendous labors in giving birth to and fostering them. Such long-term reciprocal transactions create and sustain intimate bodily and emotional ties, what Bengalis often term *maya. Maya* means attachment, affection, compassion, and love, and Bengalis think of the ties of *maya* as entailing both bodily and emotional bindings.

A range of studies has explored how, in many contexts, Indians think of persons as relatively fluid and open, that is "dividual" or divisible in nature, in contrast to the prevalent Western notion of the relatively self-contained and bounded *in*dividual. Through transactions of food, touch, objects, words, and bodily fluids, family members give and receive parts of themselves, forging not only emotional but also bodily ties.

According to such conceptualizations, by co-residing in the same household, and by giving and receiving food, material goods, bodily care, and affection, parents and descendants not only meet each other's survival needs, but also sustain intimate ties of kinship. Property and inheritance play a somewhat complicated, ambiguous role in representations of the traditional Indian intergenerational system.

Descendants are supposed to serve and provide for their elders in exchange for sacrifices their elders have already made for them. That is, elders have already earned the right to be served by their juniors, because of their tremendous earlier gifts. However, it is widely recognized that, in practice, even in the past, the expectation of an inheritance frequently serves and served as a major motivator of filial service.

Such an expectation— that those who stand to receive an inheritance from an elder are specifically obligated to provide care—becomes legally concretized in the 2007 "Maintenance and Welfare of Parents and Senior Citizens Bill". Relatedly, one reason daughters are *not* normally expected to care for their elder parents is because daughters frequently do not stand to inherit a significant degree of their parents' property at death.

Yet, even for sons, to exchange parental care for inheritance is not the "proper" way of doing things, neither in the past nor in the present. Many tales, both old and new, feature such a theme. In one classic tale, a wealthy old man preparing for his death divides his property among his sons. The sons, having inherited all they expect to inherit, promptly begin to maltreat and neglect their father. The aged man encounters an old friend and tells him of his woes. The friend promises to help.

He calls all the sons and daughters-in-law together and before them bestows to their father a large, locked, heavy chest, announcing that he is returning a long-standing debt with interest, which has now amounted to thousands of rupees. When the sons learn of their father's new wealth, they begin to be more attentive to him than ever before! The old man carefully guards the key and is well served for the rest of his days. When he passes away, the sons and daughters-in-law greedily unlock the chest only to find it filled with stones.

Importantly, the moral of such stories is never to celebrate an exchange of elder care for property, but rather to condemn the ungrateful and selfish sons and daughters-in-law who fail to recognize their normal moral obligations to care for their parents unconditionally.

I n addition to intergenerational reciprocity and *seva,* "respect for elders" is also regularly presented in daily conversations and media discourse as a vital part of Indian "tradition," as conveyed in the Minister of State's message on the International Day of Older Persons cited above: "We, the Indians, belonging to a glorious cultural tradition, have the inbuilt quality of respecting all the elders."

According to such discourses, all elders, not merely one's own senior kin, are to be respected, simply because of their seniority. Thus the director of a New Delhi–based NGO, Agewell Foundation, commented to me that if given the choice between the junior Bush and the senior Bush, Indians would never have selected the junior Bush; they would have unquestionably gone with the senior. Referring to the election of George W. Bush in 2000, he wondered why Americans had preferred the son when his father was still alive and well. He went on, speaking in English:

- If there is a seventy-eight-year-old person within an Indian family, most likely that person would be respected, although perhaps now relegated to the side due to modern pressures. In traditional India, the older person's opinions were sought before any family decisions were made, such as marriage or spending money. The senior-most person—of whatever age or medical condition—would be asked for his or her opinion and counsel.

This system of intergenerational reciprocity and elder respect is widely described in daily conversations and the public media as not only a culturally Indian, but also a natural, practical, and morally proper way of doing things.

Narayan Sarkar, a retired engineer who at age seventy lived with his wife in their south Kolkata home, their two children settled in the United States, contemplated pensively: "In our families, we raised our children—why? Our idea, our dream was that when we grew old, our sons and daughters-in-law would serve us.

And it is our dream, and a natural thing, to hope for this, to want this. We did this for our parents, and they for theirs." A middle-aged man from rural West Bengal, who lived with and cared for his incontinent and bedridden mother, spoke similarly of the system of reciprocal parent-child care as a natural and proper one, as I came upon him one morning while he was assiduously cleaning his mother's sheets:

Caring for parents is the children's duty; it is *dharma*. As parents raised their children, children will also care for their parents during their sick years, when they get old. For example, if I am old and I have a bowel movement, my son will clean it and he won't ask, "Why did you do it there?" This is what we did for him when he was young. When I am old and dying, who will take me to go pee and defecate? My children will have to do it.

The Indian press as well makes much of the "traditional Indian joint family" in almost every story it presents on contemporary aging, often contrasting the imagined past with an imperfect present: "India's grandparents were once revered and cared for by the extended families that they headed . . ."; "Famous for its culture of respect for the elderly, India is taking this tradition . . ."; "Earlier, the joint family system guaranteed care, concern and attention for the elderly . . ."; "The Great Indian Joint Family—our good old support system . . ."—narratives that I will complete shortly when turning to depictions of modernity. Deborah Moggach, in her 2004 novel *These Foolish Things,* floated the idea of outsourcing elderly populations from the developed world to India. In this playful novel, British pensioners can enjoy warm weather, mango juice with their gin, inexpensive pricing, and the Indian culture of elder care in a Bangalore retirement home.

"Her idea has a basis. Elderly care is part of our cultural values," an Indian government official is reported as remarking. I t is striking how contemporary representations so often highlight a very simple, idealized model of the joint family, one in which elders seamlessly receive loving respectful care, and no conflict, injustice, or economic hardship prevails.

However, ethnographic research has long portrayed complex tensions in the workings of joint family relations, including conflicts between mothers-in-law and daughters-in-law, elders abandoned by children in the face of poverty, youth who ridicule the old, and disjoined nuclear-style households. Moreover, although not the most dominant narrative, some of my middle-class Kolkata and U.S-based informants did describe the Indian joint family to me in more complex and ambivalent terms, as a patriarchal, restrictive, and/or conflict-prone institution, one that could be difficult especially for daughters-in-law and even young sons.

It is notable that most of the ethnographic examples I found of statements extolling the joint family come from men, while the more ambivalent statements in my field notes were often made by women.

This is not surprising, given that men have conventionally stayed within the same family from birth through death, while the position of women in both their natal and marital families is more shifting and precarious, and the difficult position of the daughter-in-law in the Indian joint family has long been a prevalent theme in public discourse. Gayatri Prasad, a middle-aged woman journalist who now lived independently in Kolkata with only her husband, her two daughters grown and married out, described the large joint family in which she had grown up.

She was the youngest of six children of her immediate parents, and they had lived with her senior uncle and his wife and children, along with her paternal grandparents. She described nostalgically how at weddings, the whole extended family would gather together for one entire month, basking in time and company, the children all having fun playing together on freshly watered lawns surrounded by flowers.

In the joint family setting, she continued approvingly, she and her siblings all learned the "core Indian values"—of family closeness, spirituality, respect for elders, the merit of taking the time to help grandfather up and down the stairs without being asked.

And she herself had learned to navigate the joint family, too, getting what she wanted, such as an education, by bypassing her own parents and going straight to her elder uncle. But she told also of how her sisters' lives had been "ruined" in the joint family, their personal desires and ambitions squelched. "Elders sometimes had *too* much power," Gayatri reflected. "They were absolute dictators.

They could be just and benevolent but not necessarily so." Two elder-abode residents and roommates, Kalyani-di and Uma-di, were chatting with me one evening about the changes under way in their society, notably the move from a joint family system to a system of nuclear families and old age homes. "Some people say that these changes are good," Kalyani-di reflected, "because a lot of arguing used to happen in large joint families." "When we were young girls, our mothersin- law would give us so much trouble! *Bapre bap!* " Uma-di interjected, and both women gasped and laughed for a moment as they recollected how they had suffered at the hands of their mothers-in-law. Then Kalyani added contemplatively, with a tone of more praise than criticism, "But, people need other people to live.

Before, no matter *how bad* relations were at home, people *had* to live together." Although in this and other conversations these women seemed to express a preference for what they perceived to be the joint family system of the past, they did not paint joint family living in simple harmonious terms. Whether viewed as glorious or fraught with tensions and inequalities, enduring or a waning relic of the past, it is difficult to stress enough how salient the "Indian joint family" is as a trope with which Indians interpret contemporary aging.

Throughout my fieldwork period, I found that a whole range of people—the old and young, male and female, those living in joint families and without, journalists and filmmakers—repeatedly expressed taken-for-granted assumptions that the most normal, familiar, and "traditional," even if not ubiquitous, way of aging and elder care for Indians is within a multi-generational or joint family.

I nterestingly, although so many present the joint family as a quintessentially *Indian* institution, they very often use the English "joint family" rather than an Indian-language equivalent, even when engaging in an otherwise Indian-language conversation.

In fact, other English terms frequently interjected into Bengali conversations —such as "culture" and "Indian culture"—are also, like "joint family," intended specifically to convey tradition or Indianness; but the use of English highlights how these concepts *become* specifically "traditional" and "Indian" in the context of and in contrast to the modern and foreign. So, as we will see over the following pages, "Indian culture" is frequently juxtaposed to "Western culture," and "joint family" to "nuclear family" and "old age home." Stacy Pigg notes how English terms are frequently used in Nepal to discuss taboo topics and mark them as modern.

But, in a related move, Bengalis also use English terms to signal concepts that have come to be viewed as quintessentially traditional or Indian precisely when they come into contrast with perceived alternative modern or Western forms.

Aging and the Forest: Late-Life Spirituality and Human Transience

In addition to using the trope of the joint family, many Indians think of traditional ways of Indian aging in terms of the Hindu textual formulation of the four stages or *aÄramas* of life. According to the classical Hindu ethical-legal texts, persons move through a series of four life stages or "shelters"—as a student, a married householder, a disengaged forest dweller and finally a wandering renouncer.

In the Hindu texts, this schema in fact applies specifically to upper-caste males, while Manu devotes little attention to defining the appropriate stages of a woman's life, which are determined by her relationships to the men upon whom she depends for support and guidance: her father in youth, her husband in marriage, and her sons in old age. But I found that both women and men, especially among the more well-educated, frequently invoked a model of the four Hindu life stages when discussing the experiences of aging. I n the *aÄrama* schema, two life phases constitute older age. When a man sees the sons of his sons and white hair on his head, he knows it is time to enter the forest-dweller or *vanaprastha* phase—departing from his home to live as a hermit, either with or without his wife, or remaining in the household but with a mind focused on God.

The final life stage is conceptualized as a time of complete renunciation of the phenomenal world and its pleasures and ties. As a *sannyasi* or renouncer, a man strives to become free from all worldly attachments, through taking leave of family members, abnegating caste identity, giving up all possessions, performing his own funeral rites, begging, and constantly moving from place to place so that no new attachments will develop. If a person is able to free himself from all binding attachments in this way, he may be able to attain ultimate "release" from the cycle of rebirths, redeaths, and reattachments to worldly life, or *samsar.*

Few Hindus actually move to the forest or become wandering renouncers in old age, but many do view late life as an appropriate and valuable time for focusing increasingly on God and spiritual awareness. They find compelling a model of the life course in which one concentrates on worldly matters—marriage, reyarn production, and material gains and pleasures—during one's adult householder years, and then turns in later life towards spirituality and consciousness of human transience.

I n some ways, though, the two models of old age explored thus far in this chapter—being served within an intimate joint family, and moving to the forest to focus on God—are in conflict, because in one a person receives care precisely within the family, and in the other the person purposefully renounces ties to the family and world.

Many of the spiritually minded older Indians I know claim to find it perfectly acceptable and possible to concentrate on spirituality in late life even while living within the family. However, as I explore in *White Saris and Sweet Mangoes,* Âothers feel conflicted and torn. They feel that their bindings of *maya*—their bodily and emotional ties to the persons, places, and things of their lived worlds—*should* decline as they grow older, but that, instead, the natural tendency is for the ties of *maya* to grow stronger and more numerous as life goes on.

They *should* be loosening ties to their families and homes, developing a readiness for death and a greater awareness of God; and yet they still experience a strong yearning to see one more granddaughter's wedding, to drink one more cup of tea from a beloved grandson's hand, to eat one more sweet mango. This theme of literal or metaphorical spiritual forest-dwelling in late life is not as salient in narratives of modernity as is the theme of the joint family, possibly because the spirituality of the aged is not so widely perceived to be threatened by contemporary social changes in India.

Narratives in which the "materialist" West is encroaching into "spiritual" India—with valuable as well as injurious results—do abound; yet many feel that Western materialism exerts even more of a pull on the young than the old. Further, some of my older informants suggest that the waning family ties of modernity and globalization may in fact facilitate latelife forest-dwelling; that several features of modern social life—such as dispersed families and even old age homes—thus might actually complement certain Indian or Hindu "traditions" of aging.

IMAGINING THE PRESENT

As those I grew to know in Kolkata and the diaspora represented what they perceived to be traditional forms of Indian aging, they also produced meanings and images about a Bengali or Indian modernity. In Kolkata, modernity is perceived as entailing a cluster of concepts and terms, including the English "modern," or *adhunik* in Bengali, and other terms conveying the temporal present, such as "these days" and "nowadays" or "now". To those in Kolkata, what I gloss as "modernity" is essentially the present itself, as different from the past— "these days" as opposed to "those days," "now" as opposed to "before."

As such, modernity is regularly associated with *Western* values, lifeways and processes, and in fact, in the nineteenth century, the Bengali term most commonly employed to describe the modern or new was explicitly linked to Western education and thought, the civilization inaugurated under English rule. Modernity today is also regularly associated with features of "globalization", such as the global spread of Western values and lifeways, a global economy and media, and transnational or diasporic living. Modernity entails as well a host of other facets of contemporary life, frequently including urban residence, nuclear families, small flats, individualism, consumerism, materialism, careerism, a persistent lack of time, weak family ties, waning patriarchy, and old age homes.

Much of what Bengalis conceive as the present was inaugurated first under foreign colonial rule and then under Western-dominated globalization; this may be one reason Bengali conceptualizations of modernity cannot but be deeply ambivalent, Partha Chatterjee argues. In contrast to the "these days" felt to be not entirely of their own making, Bengalis often construct a "'those days' when there was beauty, prosperity and a healthy sociability, and which was, above all, our own creation".

Amartya Sen, though, disagrees with Chatterjee's and other scholars' attempts to neatly distinguish and celebrate "our culture" as opposed to "their culture" prevalent in contemporary India. Instead, Sen argues that "In our heterogeneity and in our openness lies our pride, not our disgrace. Satyajit Ray taught us this, and that lesson is profoundly important for India. And for Asia, and for the world". Let us here look first at some of the stories of decline indeed so prevalent in India today.

As we have already begun to see, visions of modernity are often intertwined with images of aging, in which the most dominant theme is the decline of the joint family. One representative Englishlanguage newspaper story, reporting on the new Government of India legislation that could send children to jail for failing to look after their elderly parents, reads:

"The Indian family scenario seems to be crumbling in the nuclear age, . . . with an alarming increase in the number of senior citizens taking refuge in old age homes, . . . city parents . . . hit by the empty-nest syndrome, . . . parallel decline in respect for the elderly".

Another news story, detailing the suicide of an eighty-year-old retired Kolkata engineer whose only son lived abroad, opens: "Used to joint family structures, the elderly these days do not have even their children to talk to, leading lonely lives fraught with anguish". One Bengali gentleman reflects in a letter to the editor titled, "To Age is Now a Curse":

- During our adolescence, most Bengali households consisted of joint families. . . . the joint families or family units with one common kitchen have disappeared like dinosaurs. . . . Hence now all are nuclear families. . . . Aged parents are neglected and ignored in the family. I am now 71 plus. Most of my relatives now live in various parts of West Bengal. I have seen the miserable condition that those of my age, and even older than me, have been thrown into!17

Acclaimed Bengali film director and actress Aparna Sen, introducing a special issue titled "Problems of Aged Parents" for the popular women's magazine *Sananda,* highlights the failed reciprocity in today's intergenerational relationships: "It takes a lot of care to bring up a human baby. What kind of a human quality is it to throw into a neglected and uncared-for state those aged parents of ours without whose care it would have been impossible for us to grow up!"

The blockbuster Bollywood film, *Baghban,* one of the biggest cinematic hits in India in 2003, features the failed intergenerational reciprocity, dissolving joint family system, and chasm between the past and present envisioned as distinguishing the contemporary era.

Just like a gardener who plants trees and nourishes them in the hopes that he will be able to rest in their shade when he grows old, the movie's hero, Raj Malhotra and his wife Pooja raise a family of four sons who are all now nicely settled in their professions. Against his banker's and former employer's advice, the father upon retirement invests all of his income, including his provident fund, in his sons who, he believes, are the assets who will secure his future. "I am not worried about life after retirement at all," Raj Malhotra declares towards the film's opening.

"With God's grace we have four children. That means we have four invaluable fixed deposits. Eight hands are enough to support me." Once their father's bank account is empty, however, the sons and daughters-in-law treat the parents with contempt and refuse to care for them properly. In the end, the hero publishes a best-selling novel based on the mistreatment received from the hands of his sons; overnight he becomes millionaire, and the sons come groveling back, but he does not forgive them. The film closes as the father addresses a captive audience gathered to celebrate the book, his sons and daughters-in-law listening meekly:

- I am not a writer. . . . I have only written what life has shown me. *Baghban* is not about me or any one person. This book is about the silence that exists between the past and the future. This book is about every bridge that has broken between two successive generations. This book is about those bent shoulders on which some children sat and enjoyed the carnivals of the world. This book is about those shivering, empty hands that once held the hands of their young children and taught them how to walk. . . . The world has changed. Life has changed. People of my age please remember—

The father-author continues in a sarcastic tone, describing how the ways of the past have disappeared into the contemptible present:

- What unworthy relationships we used to be entangled in! We saw the lord's face in our father's faces. We found heaven in the feet of our mothers. But now people have smartened up. Today's generation has become very intelligent and practical. For them every relationship is like a ladder which they step on to move up in life. And when that ladder loses its worth, then like broken furniture, broken utensils, torn clothes, yesterday's newspaper, it is discarded into a storage area.

He takes on the resonant tone of an eloquent orator:

- But life does not move up like a ladder. Life grows like a tree. Parents are *not* like the first rung in a ladder. Parents are the *roots* of the tree of life. No matter how tall the tree might grow, how healthy and green it might be, cutting the roots will not allow it to remain healthy and green.

- That is why, today, I ask with great politeness and respect, that the children for whose joy a father happily spends every penny he has earned, why do these children hesitate to shine a little light for the parents when their eyes become weak? If a father can help his son to take the first step in his life, then why can't that son help his father take the last step of his life? Why are the parents, who spend their entire lives bestowing pleasures on their children, punished with tears and loneliness?

Similar images abound in my informants' own narratives, many of whom, of diverse ages and circumstances, viewed the film *Baghban* and declared it to be highly moving, truthful, and compelling. One morning I spoke with Benu, the younger son and hard-working professional of a middle-class Kolkata family, a family still arranged jointly, housing two married brothers, their wives, children, and parents.

"Today, the younger people have become very materialistic and selfish," Benu reflected. "They only think of advancing their own careers, working hard, making money, being able to buy more things. Perhaps today they are finally able to buy a car, and then they think: Why not another? It's like that." He went on to tell of how, because of the younger generation's career orientation, shortage of time, materialism, and selfishness, "Many people don't want to care for their parents any more.

They think: I'll just put them in an old age home, or live separately. Perhaps they might even get offered an apartment as a perk through their job, and they want to accept it so they can move ahead.

But perhaps their parents aren't allowed in the apartment, or there's no space for them there." He asserted that he himself, however, was totally against people "abandoning" their parents in this way and plans to live permanently as a joint family, echoing the strains of *Baghban* as he avowed: "Could a tree live without its roots? No, it would die.

My parents are my *roots*. They are the ones who raised me and nourished me and still sustain me. If they were gone, I couldn't exist—right?" I earlier introduced Kalyani-di and Uma-di, residents of a modest ten-person elder abode just a short cycle-rickshaw ride or walk from the apartment complex in which I lived with my two daughters.

One evening, as the sun was setting and we were lighting mosquito coils on the small front verandah, the conversation turned, as it so often did, to how things are so different now than in the past. Kalyani-di exclaimed: "There are *huge* transformations now in our society! Everyone has become hugely selfish! No one wants to live closely any longer. Everyone wants to live alone. You'll see—as many new homes are being built, all of them are the same kind—just for two people, two people." Uma-di interjected with a mocking smile: "Soon they'll be for just one person! Even husbands and wives will live separately!" Kalyani-di had reflected on an earlier occasion:

- "If we had grown up with the idea that we might live separately from our children, then it might not be so hard to get used to now. But with our own eyes we had never seen or known anything like this. We never could have even *dreamed* that an abode for elders existed, that we would be here, in a place like this!"

Many of the older generation comment that such dramatic modern transformations are occurring with extraordinary rapidity right in their time, so that they find themselves in the unique position of having grown up in one world and now making their lives in a radically different one. "The changes have been *sudden*! Right in *our* generation!" a Dignity Foundation member exclaimed at one of their monthly gatherings.

The Dignity Foundation is a "Senior Citizens Life Enrichment" association, founded in Mumbai in 1995 with branches in Kolkata, Chennai, and several other major Indian cities, its members mostly upper-middleclass seniors living independently from their children. Another gentleman jumped into the conversation, grinning, as he loudly voiced a lament: "We, of *this* generation, we suffered from our parents—how they rebuked us! And *now,* we suffer from our *children*—how *they* rebuke us!" However, as I have noted, visions of modernity or the present are richly multivalent. Not all are tales of unremitting disaster. A more upbeat *Sunday Times of India* story claims that the joint family—essential to Indian culture—will not disappear but is rather making a resurgence, while incorporating contemporary values: "We are finally realizing the wisdom of our conventional ways.

While nuclear families in the West go sub-nuclear in quest of the I-me-myself ideal, the Great Indian Joint Family is being reinvented. . . . Why can't we be without it? Because unlike our western counterparts, each one of us is already more than an individual: we're a network of relationships". The newly "made-over" GIJF combines older and newer elements: members have their own space but they are there for each other, too; one can call on friends as well as family for emotional fulfillment; and one can be a working woman along with a wife, mom, daughter-in-law, and aunt.

An optimistic letter to the editor from a Kolkata gentleman asserts, "A pragmatic answer to the concerns voiced would be for our society to accept and encourage the concept of old age homes as a viable replacement for the joint family of yore."

Another upbeat Sunday news story cheerfully titled "Hello! Old Age," features a group of lady friends "well into their seventies" who meet up regularly at the most modern and hip Kolkata joints, like KFC and Barista coffee. "Increasingly in Kolkata," the reporter narrates, "many senior citizens are taking charge of their own lives, wishing away dependency and redefining the conventional rules of old age."

THE SOCIAL CONTEXTS OF "MODERN AGING" IN A NEW MIDDLE CLASS

No single cause, but a whole complex of social changes, explains the emergence of novel ways of thinking about aging in India. Why *now* have old age homes suddenly burgeoned in the nation? Why now has there been a swell of institutions geared towards producing independent, non-familybased elder living? Why now has aging taken such a hold in the public imagination as a sign of the paradoxes and complexities of the contemporary age? Several interrelated factors are at work, connected to the complex reforms of the 1990s collectively known as economic liberalization; an upwelling of money from a booming economy, outsourcing jobs, and expatriate remittances; the mounting trend of transnational migration; the substantial entry of women into the professional workforce; and the intensity of India's participation in a global circulation of ideologies.

All these processes are tied to what policy makers, academics, journalists, and the public have deemed the growth of a significant new form of middle-class life in India. The processes have clustered together to create a situation where, among the urban middle classes in particular, there is more desire to create new opportunities for managing aging other than multigenerational family residence; there is more money to pay for these new options; and yet there is anxiety about what these transformations mean—for self, family, and nation. O n the level of political economy, several forces have come together to help create a burgeoning new urban middle class with noticeably more disposable income to spend and, many believe, an array of new values and lifestyles. The early 1990s in India marked a period of pronounced economic liberalization under the government led by Prime Minister Narasimha Rao.

Since achieving status as an independent nation in 1947, India had pursued economic development with limited global entanglements; but in 1991, the economy was opened significantly to foreign investment and trade. Within five years, imports doubled, exports tripled, and foreign capital investment quintupled.

Opportunities for both consumption and employment were now shaped by global markets; many professionals experienced a sharp rise in salaries; and the number of persons considered to be part of the fast-growing category of the middle class rose to somewhere between 100 and 250 million, from figures two to five times lower, over just ten years from 1989 to 1999. I n addition to bringing more disposable income into many people's hands, the new global economic climate fostered the dispersal of families across the nation and world.

With the establishment of large multinational corporations in India and booming IT, biotech, and real estate industries, many young professionals were pulled from smaller or less cosmopolitan cities such as Patna and even Kolkata to careers in global economic hubs like Bangalore, New Delhi, and Mumbai—frequently leaving behind larger multigenerational homes for nuclear-family-style flats. The past several decades have also been a period of intense transnational migration.

Although people have emigrated from India for centuries, it has only been over the past few decades that it has become commonplace, even expected, for young adults of India's cosmopolitan middle classes to migrate abroad for higher education and professional opportunities, frequently ending up settling permanently overseas.

In 2006, annual NRI remittances to India totaled over $24 billion. Indians emigrate to nations such as Australia, Canada, England, Singapore, the United Arab Emirates, and very notably the United States—where the Indian American population has risen from just about 5,000 in 1960 to over 1.5 million in 2000, a three-hundred-fold increase in just forty years. In 2002 the Government of India, recognizing the cultural and economic significance of expatriate Indians, launched a Persons of Indian Origin card scheme allowing holders visa-free entry and rights to buy property and invest in the nation, aiming as well to "reinforce their emotional bonds . . . to their original country."

The comings and goings of NRI s play a significant role in urban middle-class Indian culture, and many of the new institutions of aging are targeted specifically at the parents of Indians living abroad. One of these is the Rosedale Gardens apartment complex constructed in 2007 on Kolkata's outskirts, with 60 of the 504 apartments designed especially for the elderly living alone, and 46 of these 60 reserved in particular for senior citizens with NRI offspring—a plan conceived by two U.S.-settled NRIs who had left parents behind in Kolkata. Media globalization has accompanied economic liberalization in India, fostering as well a powerful climate of social-cultural change and heady intermingling of local and global images and ideologies.

Transnational satellite broadcasting made its Indian debut in January 1991—cable television offerings suddenly competing with state-run television, at the same time that the slackening of foreign-exchange restrictions allowed Hollywood films to vie with Bollywood yarn productions. Steve Derné notes the radical transformation of India's media:

- Fueled by advertisers trying to reach the new Indian market, the number of television channels grew from one state-run channel in 1991 to seventy cable channels in 1999. Access to television increased from less than 10 percent of the urban population in 1990 to nearly 75 percent by 1999. In 1991, cable television reached 300,000 homes;

by 1999 it reached 24 million homes. With the easing of foreignexchange restrictions, previously unavailable Hollywood films were dubbed into Hindi and screened widely.

Elaborate shopping malls, multiplex cinemas, high-rise apartment complexes with private swimming pools, restaurants, and gyms, and international food chains like McDonalds and Pizza Hut are also mushrooming across India's metropolises, almost as one watches, where nothing like that existed even just a few years earlier. India's first global-consumeroriented shopping mall opened in Mumbai in 1999, and within just five years according to one estimate, three hundred malls were under construction countrywide — replete with air conditioning, escalators, piped music, multiplex cinema theaters, food courts, and stores selling the leading Indian and international brands. "Earlier, a large majority of Indians believed in the Spartan asceticism of the Father of the Nation, Mahatma Gandhi," one journalist comments on the fast-budding Indian consumer culture.

"But the new generation of shoppers—like their contemporaries worldwide— believe in living for today and splurging at the mushrooming malls over the weekends". Although it is only the elite who can afford to shop regularly at chic malls, many less-well-to-do within the broad category of the middle class throng to gaze, watch a film, or indulge in snacks. A recent Kolkata *Telegraph* newspaper story reflects rather glibly on such changes in Bengali middle-class consumptive practices, linking them to even more pervasive transformations in value systems and ways of life:

- Call it a brand new attitude or a 21st century class struggle, but the *Bengali Middle Class* (BMC) is morphing. Just the other day, the quintessential 'Bong' was cocooned in his middle-class values, safely ensconced in tradition. Today, as a *Telegraph-MODE* survey reveals, the Bengali in Calcutta is all set for a dramatic metamorphosis.
- In almost every field—from clothes to food to social habits—the city's Bengali middle class is changing. Gone are the days when a BMC would subscribe to a life of restraint and modesty. Family axioms such as no late night parties, no alcohol, no impulsive decisions and no luxuries are being obliterated with remarkable ease.
- The statistics say it all. About 95 per cent of those surveyed agree that the BMC today is more modern and less conservative in outlook than it once was, while 42 per cent in the 20–50 age group care little about issues pertaining to morality. Eighty-two per cent feel it is important to be trendy and confident, and one out of every four polled loves to eat out.
- A casual visit to Pizza Hut or a *Kentucky Fried Chicken* (KFC) outlet on any given day would back the statistics. The swelling BMC

> crowd—which couldn't think beyond *machher jhol aar bhaat* till the other day—can be spotted tucking into crispy drums of heaven or smacking their lips after a spicy chicken and mushroom pizza.

Middle-class people in Kolkata of various ages, genders and backgrounds are talking about these matters, many finding it rather fun, new and exciting to participate in it all. At the same time, many, and not only among the older generation, engage in critical reflection, wondering if these trends are all completely desirable and if much of the Kolkata they have known and loved is fading away.

The *Telegraph* story goes on to report that a significant number of young middle-class adults has to seek mental health counseling, for as one psychiatrist explains, "When your core values receive a jolt you feel neither here nor there." And the *Telegraph* survey revealed that 91 percent in the 60–75 age group feel that the Bengali Middle Class is "going wild and losing its family values".

Finally, changing gender mores form another critical dimension of contemporary middle-class culture with an impact on aging and family life. Compared to those a generation earlier, urban middle-class women are more highly educated, older at marriage, and entering the workforce in the hundreds of thousands.

Such well-educated, professional women have significantly more agency in the family than had their mothers and mothers-in-law, creating an environment—as both older and younger persons, both women and men, concur—in which many women are simultaneously less inclined to and less able to co-reside with and care for their parents-in-law. Although I encountered some persons who lamented such changes, the more pervasive sentiment at large was that the traditional position of the daughter-in-law had always left something to be desired, and that perhaps it is not so beneficial for younger women to devote their lives towards submissively serving their in-laws in the home.

It is striking that there seems today to be less public anxiety over the movement of women from the domestic to the public sphere than over the parallel movement of elders.

The figures of old people and women are similar in that both can powerfully signify the family, upon which the "core values" of the nation hinge. In the nineteenth and early twentieth centuries, during the era of British colonialism and anti-colonial nationalism, the family was held up as essential to Indianness; yet the focus then was on women.

Partha Chatterjee examines how middleclass Bengali women in particular served to uphold a "traditional," Indian spiritual domain, located in the domestic sphere, distinguished from an increasingly "Western" materialist outer world that progressively engaged Indian men. Debates linking women with family, tradition, and Indianness have also been salient within the contemporary Indian diaspora, where Indian women are widely expected to be the critical maintainers of "Indian family values," through such practices as performing culturally attuned childrearing and community service, and preparing Indian food in the home.

Contemporary public anxieties, however, seem to center even more on the figure of the old person than on the woman. This shift has transpired as middle-class women are entering the professional workforce at a dramatic rate, taking a pivotal place in the public sphere. In fact, some current discourses represent contemporary Indian women not merely as maintainers of "Indian family values," but also as significant producers of India's new global economic success. Smitha Radhakrishnan has observed, for instance, how recent "media representations of India's booming IT industry have often carried a woman's face, suggesting that the tech revolution in India has signaled a gender revolution as well".

As the roles of women are shifting, anxieties over maintaining "core Indian values" in the home are placed increasingly onto old people and intergenerational relations. So, a cluster of events and processes have come together to foster a new kind of urban middle-class society conducive to novel ways of organizing aging and anxieties over such transitions. Some children, following career paths, have moved far from parents, across the nation or world. Some younger women are asserting preferences not to co-reside with their husband's parents.

Many parents and children feel that there is simply a different kind of "culture" at large, allowing for new lifestyles and family arrangements. There is a widespread sense, too, that a "generation gap" is increasing, partly because the pace and scope of contemporary social change seems so intense. Although of course I argued in chapter 1 that it is misleading to think that the old cannot change, partaking in new trends such as mall culture, e-mail, and gendered egalitarianism, many of my older informants do feel that the young are moving culturally forward or away from "traditional ways" faster than they are. That many of the younger generation engage in practices such as parties with alcohol, conspicuous consumption, fast food on the go instead of a sit-down rice meal, a busy lifestyle with little time for home, makes some elders uncomfortable.

So, if another option is now available, some elders conclude that they may actually prefer to live separately from their children—to keep relations easier—while still hopefully visiting and remaining emotionally close. One of the key reasons more people live outside of the conventional joint family today is also, simply, that they can afford to. That is, the establishing of separate residences and the "outsourcing" of elder care from the family to, for instance, old age homes requires economic resources that earlier generations even among the fairly well-off middle classes did not readily possess.

A BRIEF INTERLUDE: THE SPHERES OF CLASS

This chapter has emphasized the perspectives and experiences of the middle classes in India, the focus of my fieldwork for this project and of recent media attention on aging. One gets a sense from this material of a very cosmopolitan India, where people, ideas, goods, money and communications flow easily to and from a wider world. B ut I also spent some time with urban poor elders in Kolkata, especially through the *Calcutta Metropolitan Institute of Gerontology* (CMIG).

CMIG sought out the very poorest of the poor elderly in the city—those living in slums, on the streets, inside pipes, with or without their families—and invited them to come to the institute daily from noon to four, to sit under a ceiling fan, chat with each other, watch television, prepare paper packets for sale in the market, and receive food.

The institute was near my rented flat, and I would go when I could in the afternoons to hang out. The elderly women and men at CMIG interpreted their abject condition in old age not in terms of the changing ways of a global modernity but rather in terms of timeless poverty. Sons don't care for their parents because or when sons don't have money. These seniors were not participating in the complex mobilities and global interconnections that so engage their middle- and upper-class neighbours. They went to and from CMIG by foot, often stooped over while carrying a small cloth bundle of meagre belongings.

- The CMIG set on which they watched their daily hour of television was tuned to a local Bengali station. They had no kin or acquaintances living abroad.
- One sultry afternoon after we had eaten a snack of *mu®i* and tea, one among the group of ladies I was sitting with asked eagerly,
- "Say, is it true, tell me, that in the U.S., people go around wearing leaves and are otherwise naked?"
- "No, no. That's not actually true," I replied.
- "OK, another thing I heard!" she continued keenly, proposing a story that seemed to her equally far-fetched. "Is it true?—that when here it's day, there it's night?"
- "Yes, that is true."
- A nother woman joined the conversation, from her spot on the floor where she rested on a piece of burlap sack. "We've seen airplanes in the sky. They're so small! Really, can people fit in them?"

- A third woman interjected to explain that when planes are on the ground, you can see that they are very large. "Have you ever been to Dum Dum?" she asked. "If you go there, you can see how big airplanes are!"
- The other ladies were interested and asked how one gets to Kolkata's airport. I explained that, in fact, a local bus goes right from the new bypass running in front of the Calcutta Metropolitan Institute of Gerontology to the airport. It would be, I thought, less than an hour's ride.
- "How much would it cost?" one of the ladies wanted to know.
- "I think about five or ten rupees," I surmised. That would be about ten or twenty cents, or the price of a half kilo of inexpensive rice.
- "See! No wonder we can't go. Where would we get money to spend like that?"
- This simple exchange brings home just how dramatically the spatial, cultural, and economic spheres of class diverge in scale.

Bibliography

Saurabh Dube: *Stitches on Time : Colonial Textures and Postcolonial Tangles*, Oxford University Press, 2004.

Ainley, Albert: *Woolen and worsted loomfixing; a book for loomfixers and all who are interested in the production of plain and fancy worsteds and woolens*, Lawrence Mass, 1900.

Alford, Marianne Margaret Compton Cust: *Needlework as art*, London: Sampson Low Marston Searle and Rivington, 1886.

Alfred Jenks & Son: *Illustrated Catalogue of Machines Built by Alfred Jenks & Son, Manufacturers of Every Variety of Cotton and Wool Carding, Spinning, and Weaving Machinery, in All Its Departments*, Bridesburg, Pa.: Alfred Jenks & Son, 1853.

Allen, Frederick James: *The shoe industry*, Boston: The Vocation bureau of Boston, 1916.

Allen, Nellie B: *Cotton and other useful fibers*, Boston: Ginn, 1929.

Allington, Sara May: *Practical sewing and dressmaking*, Boston: Page Co., 1913.

Allinson, May: *Dressmaking as a trade for women in Massachusetts*, Washington: Govt. print. off., 1916.

Atkinson, Edward: *Cotton and Cotton Manufactures in the United States*, Boston: Franklin Press, 1880.

Baird, Robert H: *The American Cotton Spinner, and Managers' and Carders' Guide: A Practical Treatise on Cotton Spinning*, Philadelphia: H. C. Baird, 1869.

Batchelder, Samuel: *Introduction and Early Progress of the Cotton Manufacture in the United States*, Boston: Little, Brown and Company, 1863.

Brown Brothers & Co: *Brown Brothers & Co., Successors to Butler, Brown & Co.: Manufacturers of Shaw's U. S. Standard Ring Travelers, Belt Hooks, Wire Goods, Etc., and Dealers in Supplies for Cotton, Woolen, Silk, Jute & Flax Mills, Providence*, R.I.: Brown Brothers & Co., c1881.

E. Jenckes Manufacturing Co: *Illustrated Price List: E. Jenckes Mfg. Co., Sole*

Manufacturers of Hicks' U. S. Standard Ring Travelers, Manufacturers' Supplies of Every Description, Pawtucket, R.I.: E. Jenckes Mfg. Co., 1883.

Lowell Machine Shop: *The Lowell Machine Shop, Lowell, Mass.: Builders of Cotton Machinery of Every Description*, Lowell, Mass.: Stone, Bacheller & Livingston, 1882.

Mason Machine Works: *The Mason Machine Works, Taunton Massachusetts, U.S.A.: Inventors and Builders of Cotton Machinery*, Taunton, Mass.: The Company, 1898.

Merrill, Gilbert R: *American Cotton Handbook: A Practical Reference Book for the Entire Cotton Industry*, New York: American Cotton Handbook Co., c1941.

Scott, Robert: *The Practical Cotton Spinner, and Manufacturer: The Managers', Overlookers' and Mechanics' Companion*, Corr. and enl. Philadelphia: H. C. Baird, 1851.

Webber, Samuel: *Manual of Power for Machines, Shafts, and Belts, with the History of Cotton Manufacture in the United States*, New York: D. Appleton and Company, 1879

Index